HyperTaxonomy
- a computer tool for revisional work

F. Skov

AAU REPORTS 26

Botanical Institute Aarhus University 1990

CONTENTS

Preface

This book is a manual and a users guide to HyperTaxonomy, a computer program designed to assist herbarium taxonomists in the process of writing taxonomic revisions or floras.

HyperTaxonomy is a HyperCard™ application for Macintosh™ computers. My aim was to design a taxonomic work station which could manage most data needed for herbarium taxonomy and taxonomic revisions, including text, figures, maps, and drawings. In order to make the program easy to use, screen forms were made to simulate the familiar card indices commonly used by most taxonomists.

HyperTaxonomy creates a number of reports in text file format, including full taxonomic reports with complete nomenclature, and literature- and specimen citations.

The program handles morphological data and generate descriptions automatically from raw descriptive data. A multi access key allows the user to identify collections and check descriptions.

Distribution maps are available for all taxa included in a study. They can be viewed on-screen or printed.

HyperTaxonomy also includes dictionaries with abbreviations of periodicals and author names.

Flemming Skov
Aarhus, August 1990

Acknowledgements

I am indebted to Susanne Renner and Bente Eriksen who tested early versions of HyperTaxonomy and suggested many valuable changes. Henrik Balslev read and commented the manuscript.

Author

Flemming Skov. *Born 1958. Cand. scient. Aarhus University 1986. Ph.D. Aarhus University 1990. Since 1989 employed by the Danish Research Service for Plant and Soil Science, Research Centre Foulum, P.O. Box 23, DK-8830 Tjele, Denmark*

Author

[illegible]

1. INTRODUCTION

Leenhouts' guide to herbarium taxonomy (1968) has traditionally served as a guide for revisional work. It is based on years of accumulated experience in the practice of taxonomic revisions and represents a well tested method, suitable for small as well as large taxonomic studies.

Since 1968, however, the world has experienced a revolution in computer development. Computers have become faster, more powerful and, in the same period, less expensive. Most taxonomists and institutions can afford computers with adequate power for managing taxonomic data.

Development of suitable software for herbarium taxonomists has not kept up with the hardware development. Much of the tedious routine of taxonomic work is done much faster with a computer, but computer programs have been too difficult to use or have had too restricted potentials. There are no programs specially developed for herbarium taxonomy, and most taxonomists prefer to use standard software and squeeze taxonomic data into a relational database or a spread sheet package. Many of these applications are adequate for one type of data, but taxonomists need at least four different programs to manage all their data:

1. A relational data base (for managing data relations between collections, literature, and taxonomic names)
2. A visual data base (for the construction of distribution maps)
3. A spread sheet package (for managing morphological measurements of collections and taxa)
4. A word processor to write and edit the manuscript.

Using too many applications for one task is asking for trouble in the world of computers. Transference of data between applications is one problem. It is difficult and very time consuming. Updating related data is another complication: if a determination of a collection has to be changed, it must be changed in all four applications. This takes time and introduces errors. A user friendly computer program which could manage all data necessary in a taxonomic study was therefore needed.

Computers and taxonomy

The first attempts to computerize herbarium routines dates back to the mid-sixties, where a few centres were experimenting with the creation and management of curatorial databases (Gómez-Pompa *et al.*, 1985).

Computers at that time were large, expensive, and relatively inefficient, and the development of faster and more powerful micro–computers and better software has revolutionized computing in many scientific establishments. Allkin (1989) noted that existing database software is neither suitable nor easily adapted to the needs of taxonomist and that even programs written specifically for taxonomic purposes are too difficult to use for many taxonomists. Examples of taxonomical databases written with commercially available database software are found in Pankhurst (1988), and Beaman and Regalado (1989).

Allkin (1989) describes the ideal "taxonomically intelligent program" as the sum of 1) a taxonomic data structure, 2) algorithms to implement taxonomic procedural rules, and 3) a taxonomic interface to hide the underlying complexity of the data base.

Data structure

Pankhurst (1988) divided the data needed for a monographical or floristic project into four broad categories:

- Curatorial (specimens)
- Nomenclatural (names)
- Bibliographical (literature)
- Morphological (descriptions).

Most taxonomical database projects include most of these categories, but differences arise if the purpose of the database is mainly specimen-oriented or taxon-oriented (Beaman and Regalado, 1989).

Data can be structured or stored as free text. Most authors recommend that databases are as structured as possible (Allkin and Bisby, 1988). A structured database permits automatic checks of data integrity, and provides for flexible indexing and reliable retrieval, but imposes many restrictions on a user, because his data may not fit into a rigid frame. In my view, a balanced use of free-text entries and structured fields is the best solution.

Descriptive data are the most difficult to handle for a computer. A program must be able to manage the taxonomic hierarchy, logical dependencies among characters, and taxonomic variability (Allkin and Bisby, 1988). Most programs for handling taxonomic descriptions, are based on the DELTA format, originally designed by Dallwitz (1980).

Taxonomic algorithms

Taxonomic algorithms are routines undertaken by the computer to perform taxonomic operations. Examples are numerous, *e.g.*, the construction of lists of collections of a given taxon, the construction of a synonym list, retrieval of data about a literature reference, an automatic description of a species, *etc.* Allkin (1989) stresses that the "intelligence" incorporated into taxonomic algorithms not is capable of taxonomic analysis, but rather act as a complement to the taxonomists own judgement.

Taxonomic interface

Allkin (1989) defined a taxonomic interface as a "user interface to hide the underlying complexity of the database while allowing taxonomists to manipulate their data as flexible as possible using terminology and concepts already familiar to them". A well developed user interface is often neglected by programmers partly because most computers have limited graphic potentials. An extended use of metaphors in the design of screen forms would help many taxonomists to overcome their computer aversions and benefit from some of the obvious advantages from using them.

HyperTaxonomy

In 1987 Apple released a program called HyperCard. It is a computer tool for developing applications of all kinds. Its main asset is its extreme flexibility combined with a very user-friendly interface. HyperTaxonomy was developed in 1988 with HyperCard version 1.2. My aim was to construct a program which incorporated:

Leenhouts' three indices

- An index to all collections in a study
- An index to the literature
- An index to the taxonomic names.

Beyond Leenhouts

- A file containing measurements of collections

- Distribution maps of species
- A dictionary with commonly used abbreviations of periodicals and author names.

The program should be able to

- Assist the user during data entry and data treatment
- Organize the data logically and according to taxonomic practice
- Write full reports in text file format
- Produce automatic descriptions of species using the raw morphological data
- Produce distribution maps
- Identify collections with a multi access key.

How to use this manual

The basic properties of HyperCard are described in the next chapter. Following chapters describe the stacks included in HyperTaxonomy. Most chapters have these entries:

- Card design — a short guide to the fields of the stack
- Adding and deleting cards
- Entering information — how and where to enter data
- Actions — what to do with a stack - data manipulation
- Navigation — where to go from a stack.

2. HYPERCARD ESSENTIALS

This chapter deals briefly with the basic properties of HyperCard. The references listed may be consulted for an extended discussion of the subject.

Cards, stacks, backgrounds, and fields

The basic unit in HyperCard is the card. A card is what is actually shown on the screen. Cards have a fixed size corresponding to the exact size of a Macintosh SE or Macintosh Plus screen. A card is analogous to the paper index card traditionally used for taxonomic revisions.

Collections of cards are called stacks. A stack can be compared to a card index or a card file, *i.e.*, a drawer filled with cards holding information about a specific subject.

The three main stacks in HyperTaxonomy will be familiar to taxonomists: one is for information about collections, one for literature references, and one is for taxonomic names.

Other stacks can not be compared to a simple card index, they are more like programs than files and are used for indexing, reporting, and overall control. Each stack is represented as a HyperCard document icon on the Finder (see Fig. 5).

A stack has at least one background. A background contains graphic elements, fields and buttons common to all cards of that particular background.

Text information is stored in fields. The total length of text allowed on one card is 32,000 characters. This is equivalent to ca. 15 A-4 pages of text. The number of cards allowed in a single stacks is limited by available disk space only.

Buttons

Clicking on a button in a HyperTaxonomy stack will cause something to happen: show another card, get information, show one field and hide another, draw a graph, or print a report.

Buttons have many forms, many of which are familiar to Macintosh users. Others are designed especially for HyperCard. Buttons are sometimes invisible or rather blended in with the background. Try to press the Command and Option keys at the same time. All buttons become visible. Some common button types are shown in Fig. 1.

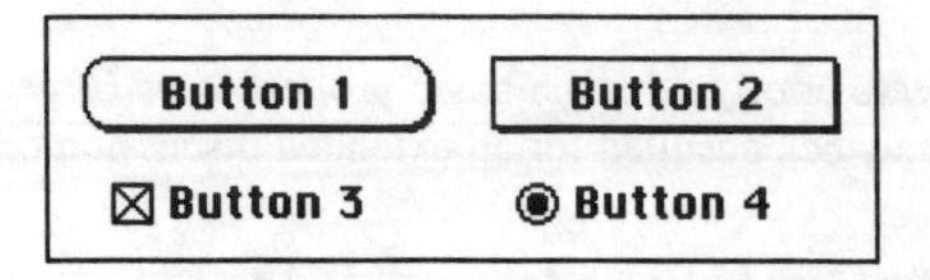

Figure 1. Common button types in HyperCard.

Cursors

Cursor are of several shapes:

- Browse tool cursor (a small hand). This is used for button pressing. In user level 1 the cursor will always be in the browse tool mode.
- I-Beam cursor (text editor). The cursor takes the shape of an I-Beam cursor when placed atop a text field. It indicates, that text can be entered or edited.
- Watch cursor. This cursor indicates that some action is going on and the user must wait until the program finishes.
- Beach-ball cursor. This cursor is used by HyperCard and indicates a sort or a search.

Message box

The message box can be used to communicate messages between the user and HyperCard. Too see the message box, type Command-M or choose Message in the Go menu. HyperCard "speaks" in a language called HyperTalk (see manuals for further details). One line can be entered. Press Return to execute the command. The most common use of the message box is for "go", "find", or "sort" commands. See examples below.

The message box can also be used as a small calculator. Type a formula and press Return to see the result (Fig. 2).

Navigating in HyperCard

Home button

All cards have a home button in the lower left corner. The home button is iconized as a compass rose (see Fig. 3, left). Pressing a home button will show the first card of the home stack.

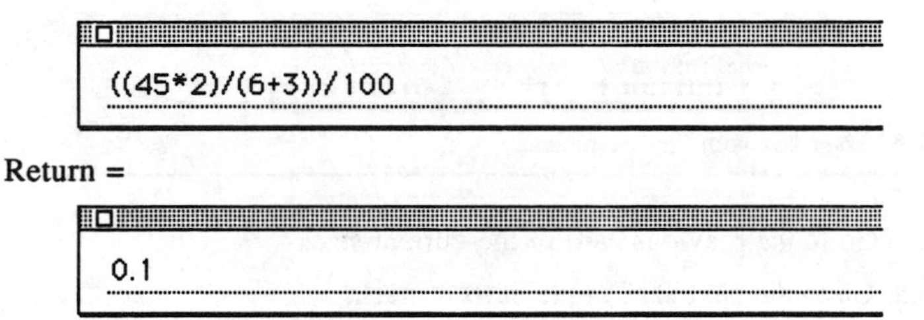

Figure 2. The message box used as a calculator.

Arrow buttons

Most stacks have navigation buttons. Fig. 3 shows three types of arrow buttons.

- Previous button (left arrow). Go to the previous card in the stack.
- Next button (right arrow). Go to the next card in the stack.
- Return button (return arrow). Use this button to go to the previous card in another stack.

Go menu

The Go menu has the following options:

- Back. The back command takes the user to the most recent card.
- Home. Takes the user to the first card in the home stack.
- Recent. Lets the user backtrack any of the last 42 different cards seen. When a card is displayed, HyperCard puts a miniature picture of it on Recent. To go back to a specific card, just click that card's picture.
- First. Go to the first card in the current stack.

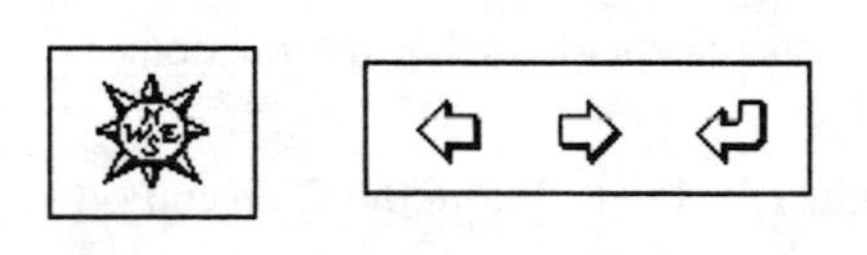

Figure 3. Home button and three navigation buttons.

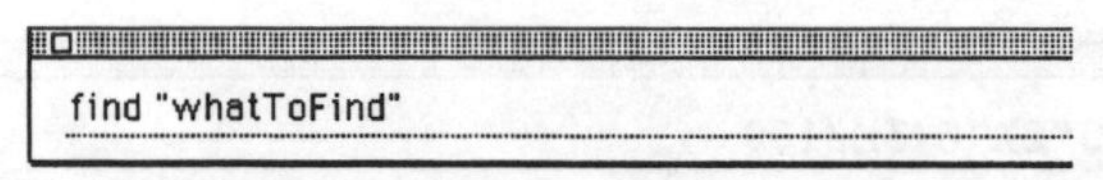

Figure 4. Message box with Find command.

- Prev. Go to the previous card in the current stack.
- Next. Go to the next card in the current stack.
- Last. Go to the last card in the current stack.
- Find. Brings up the Message box, with the Find command typed in.

Navigating with the message box

The message box can also be used to navigate in HyperCard
Use the Go command:

- "go to the first card of stack collections"
- "go to card 125 of stack taxonomy"
- "go home".

Keyboard arrows

Macintosh SE and Macintosh II with extended keyboards have four arrow keys which can be used for navigating.

- Left arrow. Go to the previous card in the current stack.
- Right arrow. Go to the next card in the current stack.
- Down arrow. Go to the last card visited. HyperCard remembers the last 100 cards visited.
- Up arrow. Reverse the order of the Down arrow.

Find

Most screen forms have a FIND button. Pressing this button will display the message box as shown in Fig. 4.

There are several other ways of finding information:

Using the Go menu

- Choose menu item "Find..." from the Go menu

  or

- Type Command-F.

Using the message box

- Choose Message from the Go menu
- Type: *Find "whatToFind"* and press return.

The Find command can be extended by adding one of the commands shown in Table 1. The search can be narrowed to a single field: to find "*Geonoma acaulis*" in the title field of the literature stack, type the following into the Message Box: find whole "*Geonoma acaulis*" in field "title".

When HyperCard finds a text string it is marked with a box. If a card is displayed without a box around any of the words, the word appears in a field not currently displayed.

> NOTE: The Find command works only for the current stack.

Sort

Stacks for storing data have a SORT button with some predefined mode of sorting built-in. Stacks can, however, be sorted by all fields or field components.

Sort using the Message box

- Choose Message from the Go menu
- The general sort command looks like this:

  sort {ascending/descending} {numeric/alphabetical} by <expression>

Table 1. Use of the Find command in HyperCard

Command	String without spaces	String with spaces
find " "	at beginning of word	multiple search
find chars " "	anywhere inside word	multiple search
find word " "	whole word only	multiple search
find string " "	anywhere inside word	single search (contiguous string)
find whole " "	whole word only	single search (whole words only)

- The expression argument must be a text expression which holds a field component address.

Examples:
sort descending numeric by field "collector"
sort ascending by word 2 of field "locality".

Deleting and adding cards

It is possible to add new cards or delete existing ones in all stacks by choosing "Delete card" or "New card" from the Edit menu. Use this feature in collections-, B-P-H-, and maps stacks only.

User level

HyperCard stacks can be set to user levels between 1 and 5.
Higher user levels make additional tools and powers available to the user.

1. Browsing. No text entry possible.
2. Editing. Text entry possibly.
3. Drawing. The drawing tools can be used.
4. Authoring. The button tool and the field tool can be used to move, resize, and delete buttons and fields.
5. Scripting. HyperTalk scripts can be read and edited.

How to set the user level

There are two ways of setting the user level to a specific level.

- Go to card "User preferences" of stack home (see chapter 3)
- Choose the radio button corresponding to the desired user level

  or

- Select Message from the Go Menu
- Type "set userLevel to 2"
- Press Return.

3. HYPERTAXONOMY

What is needed

To use HyperTaxonomy you need the following software:

- HyperCard® version 1.2
- Apple®system software version 4 or later.

HyperTaxonomy includes:

- A home stack (Home)
- Collections stack (Collections)
- Literature stack (Literature)
- Taxonomy stack (Taxonomy)
- Stack with distribution maps (Maps)
- Stack with abbreviations (B-P-H)
- Spread sheet stack with morphological data (Measurements).

One of the following three configurations of hardware in addition to the above mentioned software is needed:

- A Macintosh® Plus with a hard disk
- A Macintosh® SE or SE/30 with a hard disk
- A Macintosh® II, IIx, IIcx, IIci, or IIfx.

To print documents you will need an Apple ImageWriter® or an Apple LaserWriter®.

How to start with HyperTaxonomy

Before starting the program, check if you have everything you need.
Make a backup copy of the program disk on the hard disk to work with.

> IMPORTANT: Keep HyperTaxonomy stacks and the HyperCard application in the same folder in case you have more than one home stack on your hard drive. HyperCard always goes to the nearest home stack when it is opened.

Fig. 5 shows the HyperCard application icon and all stacks used in HyperTaxonomy as they appear on the desktop.

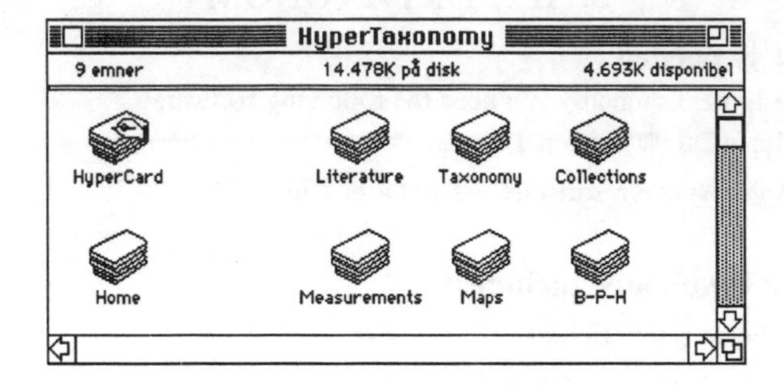

Figure 5. HyperCard application and stacks used in HyperTaxonomy as they appear in the Finder.

Double-click on the HyperCard program icon or on the home stack icon on the desktop to start the program.

User preferences

Press button User Prefs on the home card (see Fig. 7) to show the card where user preferences are set.

This card is a small control panel to HyperTaxonomy where parameters

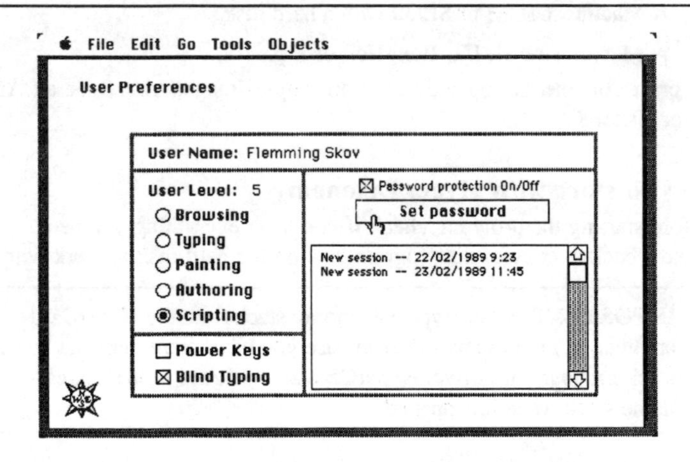

Figure 6. Card for setting the User preferences.

like user level, blind typing (*i.e.*, the ability of typing something into the message box when it is not displayed), and power keys (special key strokes which are useful when operating in the paint mode) are set.

Password
Stacks can be password protected. When password protection is on HyperTaxonomy asks for the password when it is launched from the Finder.

How to set the password

- Click button Set Password
- Enter the new password two times
- If the two words are not identical, you must try again
- Check button Password on/off if you want to use the password protection. When HyperTaxonomy has been opened with the password protection, button Password on/off is automatically set to off to prevent unnecessary use of the password dialog box.

4. HOME STACK

The home stack is the centre of HyperTaxonomy. Its main purpose is control and not data storing. The home stack controls:

- Access to all other stacks of HyperTaxonomy
- Reporting
- Updating of linked files
- A key to the taxa in the study
- Access to other applications on the Macintosh (Minifinder)
- Cards with user preferences and pathways to stacks, applications and documents.

> NOTE: It is not possible (nor desirable) to add or delete cards in the home stack! The New Card and Delete Card options from the Edit menu does not work in the home stack.

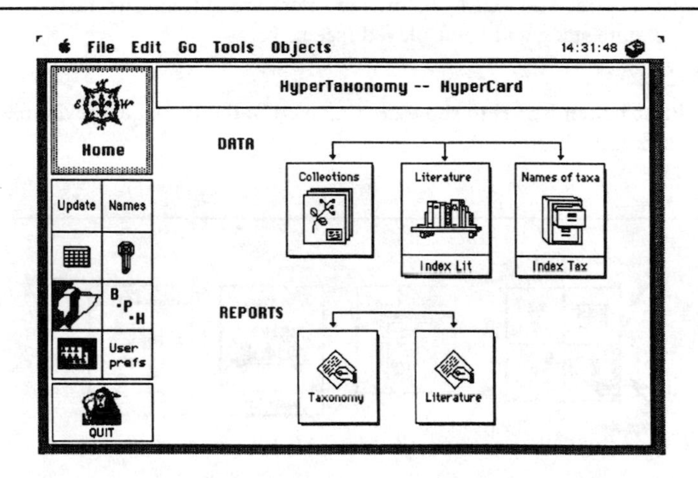

Figure 7. The home card - centre of HyperTaxonomy.

Home card

The home card is the first card shown when a HyperTaxonomy session begins. It is a starting point from where all parts of the program can be reached. To go to the home card from anywhere in HyperTaxonomy press the home button, which is symbolized as a compass rose, or press Command-H. The home card is shown in Fig. 7.

Access to main stacks (Fig. 8)

Collections stack

- Press button Collections. HyperTaxonomy shows the first card in the collections stack (see chapter 5).

Taxonomy stack

- Press button Names of taxa. HyperTaxonomy shows the first card in the taxonomy stack (see chapter 7).
- Press button "Index Tax" below the Names of taxa button. The index to the taxonomy stack will be displayed (see p. 17).

Literature stack

- Press button Literature. HyperTaxonomy shows the first card in the literature stack (see chapter 6).
- Press button "Index Lit" below the Literature button. The index to the literature stack will be displayed (see p. 17).

B-P-H stack

- Press button B-P-H to show the first card in the B-P-H abbreviations stack (see chapter 10).

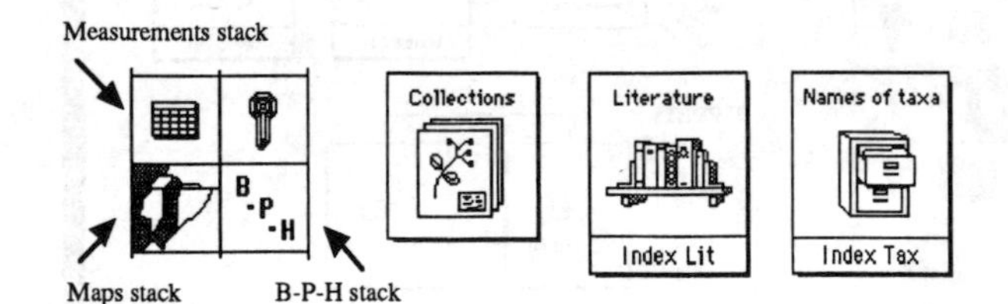

Figure 8. Navigation buttons from the home card give access to main stacks.

Maps stack

- Press this button to see the first card in the maps stack (see chapter 9).

Measurements stacks

- Press this button to see the first card in the measurements stack (see chapter 8).

Indices to literature and taxonomy

Indices are provided to all references in the literature stacks and all names in the taxonomy stack. For each index, a header indicates the number of entries available to the user.

Index to taxonomy stack stack

HyperTaxonomy offers three different indices of the taxonomy stack:

- All names in the taxonomy stack including synonyms, dubious, and illegal names
- All accepted names (= all valid names, not synonyms nor illegal names)
- An index for common taxa where collections have been recorded in the study area.

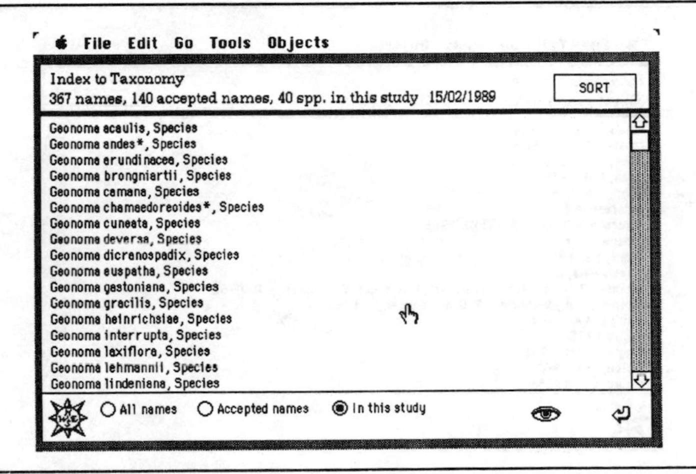

Figure 9. Index to taxonomic names.

The index card is shown in Fig. 9.

How to use the index

- Select an index with one of the three radio buttons below
- Find the desired name in the index (use the scroll bar on the lower right to bring any part of the index into view)
- Point and click at the name
- HyperTaxonomy displays the selected card in the taxonomy stack.

How to sort the names in the index

- Click button Sort
- Choose ascending or descending sort order.

> NOTE: Sorting the names in the index does not change the order in which the cards appear in the stack.

How to edit an index

Errors in an index can be corrected: a click on the "Eye" button at the lower right permits editing of the list.

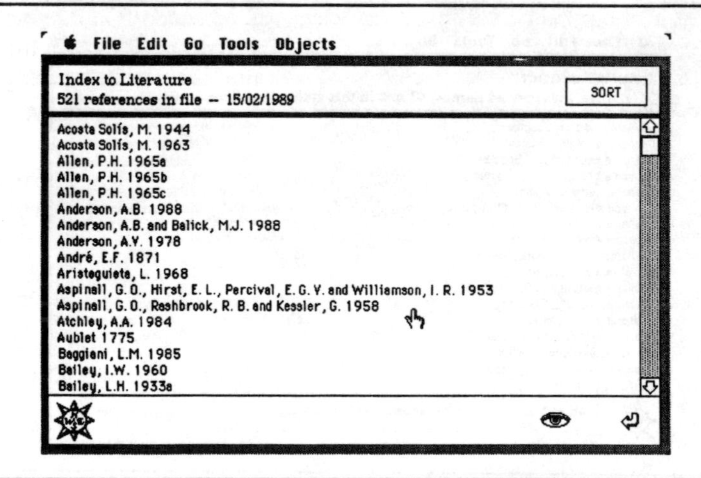

Figure 10. Index to the literature.

> WARNING: Be careful not to change correct entries: each line holds a key word. If anything goes wrong: rebuild the index (see p. 24).

Index to the literature stack
The index to the literature is shown in Fig. 10. For a discussion of the use of buttons see above.

Report section

HyperTaxonomy produces text reports in ASCII text file format. This format is easy to transfer to a word processor. The program supports literature reports (= reference lists) and taxonomic reports.

Both types of reports require the same basic steps:

1. Choose which type of report to work with
2. Make a list of items to be included in the final report
3. Choose the format in which the report should be made in.

Press the Taxonomy or the Literature button on the home card according to the purpose of the report (Fig. 11).

A card containing a list of items which can be included in a report is displayed (Fig. 12). The card has two fields. The upper field holds a list of all the names included in the study. The lower field is for building a list of items which are to be included in the final report.

Figure 11. Buttons on home card for taxonomy- or literature report section.

Taxonomy report

Note that a choice between all taxa names, and the taxa represented in the study area with collections is available (see Fig. 12).

How to prepare the list

- Press button Clear All to clear the contents of the lower field
- Press Copy All if you want to include everything in your report or click on the first name to be included in the report
- Continue until the list is complete.

Sorting the list

The list is build in the order in which the names are chosen, but can be sorted afterwards:

- Click button Sort and choose ascending or descending sort order in the dialog box.

Delete items from the export list

The whole list can be cleared by clicking button Clear All. A single item can be deleted by pointing and clicking on it.

Literature report

The choice card for the literature stack is shown in Fig. 13. If you want to

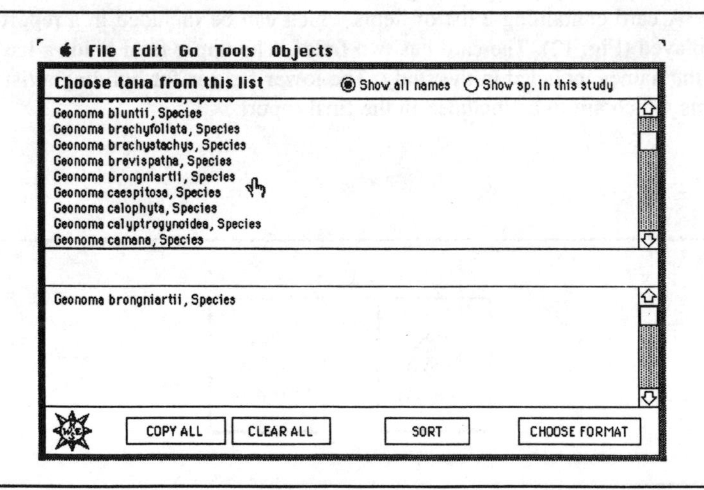

Figure 12. Use this card to build a list of names to be included in a taxonomic report.

see the card of a reference instead of copying it to the list press button Copy. This action change the button icon to an eye. Clicking on an item in the list will now show the card. Press the mouse button once to return. Press the "eye" button again to switch back to copy mode. This ability is often useful because it is difficult to remember what is" behind" a long list of references.

Formatting

After a list is completed click Choose Format to continue. This button shows the taxonomic format card or the literature format card.

Taxonomic formats

HyperTaxonomy supports six different formats (see Fig. 14):

- Flora of Ecuador format
- Flora Neotropica format
- Draft/overview format
- Exsiccatae (for all taxa with species- or lower rank in the study)
- Morphological data in DELTA format (not fully implemented!)
- Morphological data for analyzing data and drawing graphs.

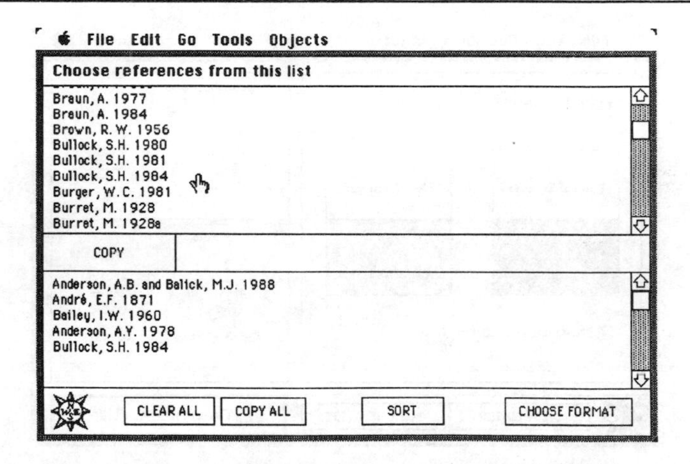

Figure 13. Use this card to create a list of references to be used in a report.

Checking information
If the taxonomy stack is not fully updated, press the Check information button. HyperTaxonomy will check and update all links to other stacks during report construction.

How to edit the order in which provinces or countries appear
In certain formats specimens are sorted after province or country, which in turn, are sorted in some specific way. Pressing the Show Editor button displays a field, where any order of provinces/countries can be entered (Neotropica versions of HyperTaxonomy, *e.g.*, uses countries). See Fig. 15.

Apply taxonomic status to each name in the report
If button Show Status is checked, HyperTaxonomy writes the status of each taxon (except for accepted names) after the taxon in the report. Remember that HyperTaxonomy lists synonyms in strict chronological order. The status of each taxon can help you to rearrange the taxa in homotypical blocks in a word processor afterwards.

Changing the list of names in the report
Choose button Make New List. This button brings back the choice card discussed above where the list can be changed.

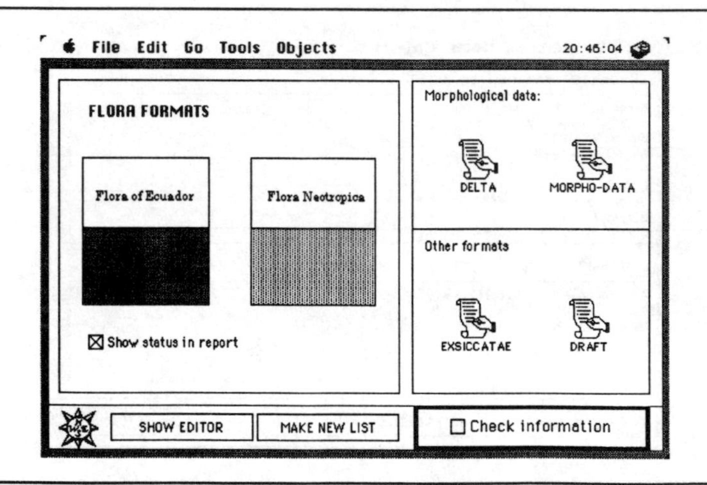

Figure 14. Output formats available for the Taxonomy section in HyperTaxonomy.

How to create a report

- Press the button with the desired format
- A dialog box will ask you to type a name for the text file, which is going to be the result of the report
- Click OK after naming the report
- Wait.

Literature formats

HyperTaxonomy supports four different literature formats following the instructions for contributors to these periodicals (Fig. 16).

- Nordic Journal of Botany format
- Nature format
- Taxon format
- Journal of Ecology format.

Changing the list of items in the list

Press button New List. This button brings back the choice card, where the list can be edited.

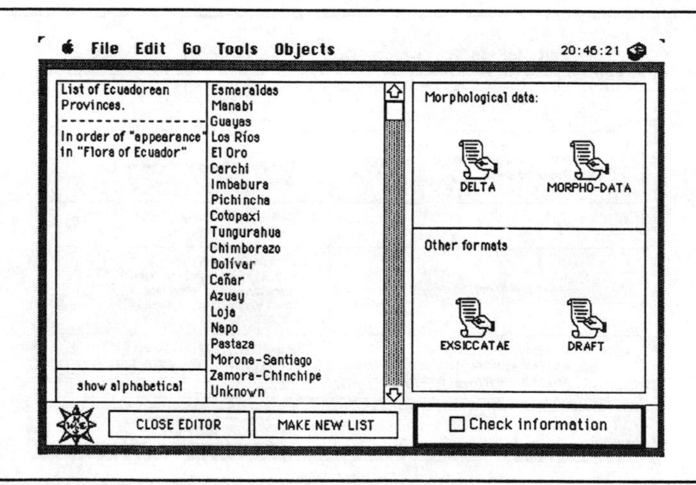

Figure 15. Field where the order of provinces can be edited.

How to create a report

- Press the button with the desired format
- A dialog box will ask you to type in a name for the text file, which is the result of the report
- Click OK after naming the report
- Wait.

Update stacks and indices

This card controls "long distance" updating of indices and the taxonomy stack (Fig. 17). Indices must be updated from time to time if the update index option has not been used when entering cards or changing key fields in the literature and taxonomy stacks. Changes in the synonomy and new species found in the study area likewise call for an updating of the taxonomy index. The taxonomy stack needs updating when information from the collections- and the literature stacks no longer correspond.

How to construct new indices

- Decide which indices to update

File Edit Go Tools Objects 9:22:45

1.	Nordic Journal of Botany Opera Botanica	Balslev, H. & Barfod, A. 1987. Ecuadorean Palms - An overview. - Opera Bot. 92: 17 - 35.
2.	Nature	Gentry, A.H. Pl. Syst. Evolut. 137, 95-105 (1981)
3.	Journal of Ecology	Chazdon, R.L. & Fetcher, N. (1984). Photosynthetic light environments in a tropical lowland rainforest. J. Ecol, 72, 553-564
4.	Taxon American Naturalist American Journal of Botany	Burret, M. 1929. Die Gattung Euterpe Gaertn. Bot. Jahrb. Syst. 63: 49-76.

New List

Figure 16. Formats available for the literature section in HyperTaxonomy.

- Check one, two, or all four check buttons
- Click OK and wait.

How to update the taxonomy stack with linked information
Update options are similar to the ones present in the Links menu in the taxonomy stack (see chapter 7).

- Check which links to be updated
- Before clicking OK, select one of the following options:
- Update all cards in the taxonomy stack
- Update a range of cards only (select start and end number)
- Update the cards included in the special index.

How to update the measurements stack
Use this button to update all cards in the measurements stack. It merges the data and then creates new descriptions for all taxonomic entities of species level and above (see also measurements stack, chapter 8).

How to change the taxonomy

Use this card to change a name (and all its synonyms) into a synonym of

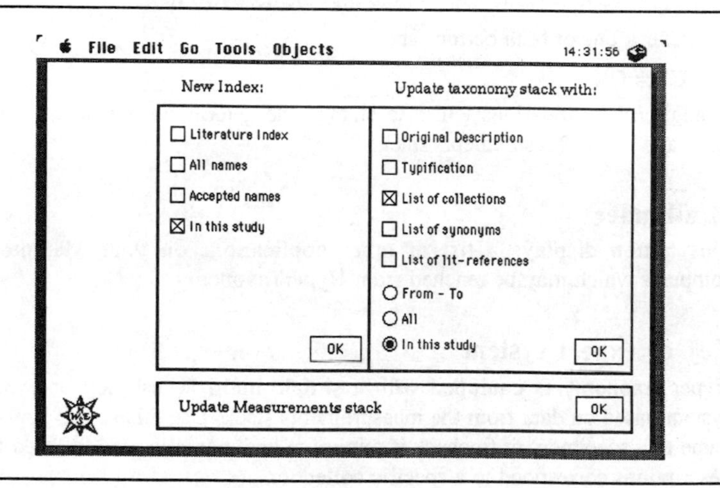

Figure 17. Update indices and stacks.

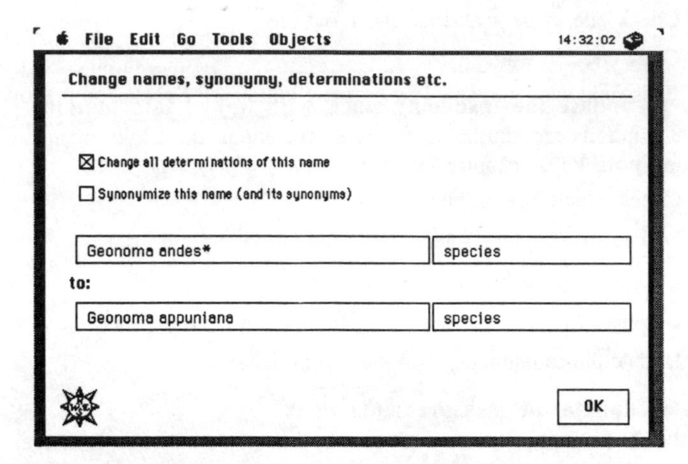

Figure 18. Card for changing names and determinations.

another name, or to change all determinations of all collections of a given taxon to another name (Fig. 18).

- Type the name to be changed and its rank into the first two fields
- Type the new name and its rank into the next two fields
- Check one or both buttons above
- Click OK.

Changing determinations will also change the determinations in the maps stack and in the measurements stack.

MiniFinder

This button displays a list of other applications on your Macintosh computer which may be reached from HyperTaxonomy.

Key or expert system

HyperTaxonomy is equipped with a simple multi access key or expert system based on data from the measurements stack. Use this key to apply a name to a specimen, or to check if minimum and maximum values used for descriptions correspond to a specific collection.

Press the Key button on the home card to see the card shown in Fig. 19. The card has four fields with different purposes:

- A field containing a list of all taxa present in the measurements stack. This field can be updated by pressing button All Species on the top of the card.
- A "positive" field for a list of the species names you think it might be (in the beginning of a search this field holds all names, unless you want to exclude some of them before you start).
- A "negative" list. This field is empty before you start your search. After a search it contains the taxa which are excluded by HyperTaxonomy.
- A field with your search criteria.

How to set the search criteria

Press button Set Criteria to choose the character you want to enter. Values for quantitative characters are entered as numbers. If a qualitative character is chosen, its character states will be shown on the screen. Fig. 20 shows the sequence. Continue until all desired characters have been chosen. Note that the new setting has been entered in the lower field. To delete the current search criteria press Clear.

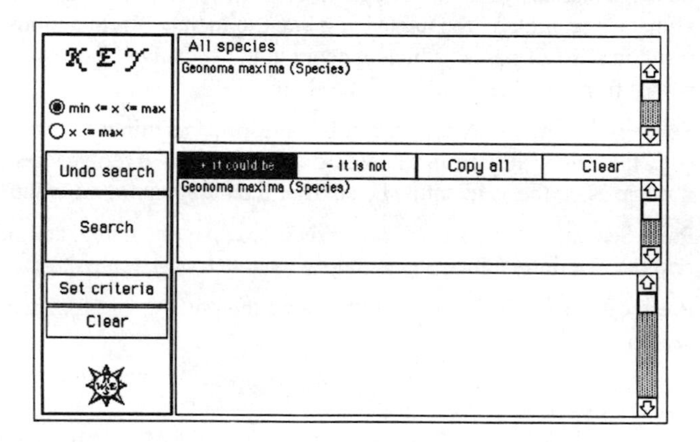

Figure 19. Key or "Expert-system" to identify taxa.

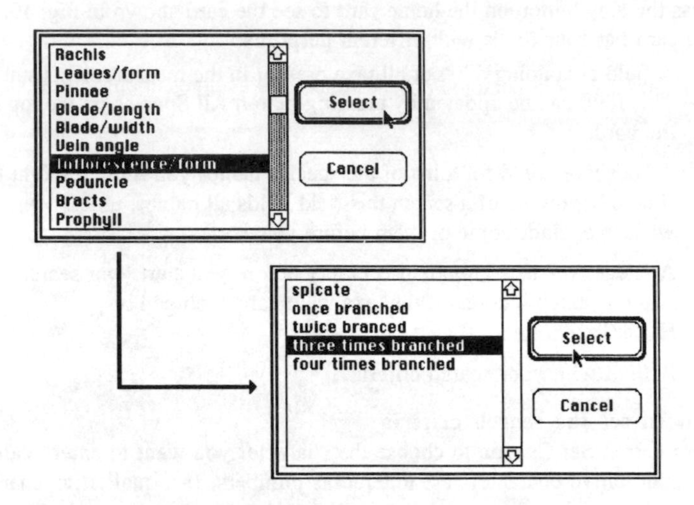

Figure 20. How to select a qualitative character as search criterion.

How to perform a search

- Select the names you want to examine. Click >> if you want to search all names or click on the names you want to include. Both actions result in a list of names which appears in the "positive" list. Exclude names from this list by clicking on them.
- Select search mode. A search can be performed as min < x > max (*i.e.*, the value must be in the recorded range for a given species), or x < max (*i.e.*, the value must be smaller than the maximum value
- Press Search. If the positive list holds more than one name, add new criteria. Continue until you can apply a name to a specimen.
- Press Undo search if you want to restore the positive list for the last search.

5. COLLECTIONS STACK

This stack contains information about collections. Most data come from specimen labels, but personal observations such as notes about morphology can be added. A small map in the upper right corner of the card shows where in the study area the collection is found.

Card design

A card from the collections stack is shown below.
A card contains the following fields:

1. Collector (main)
2. Collection number
3. Collected with (additional persons listed as collectors on the label)
4. Country
5. Date
6. Province
7. Altitude
8. Determination

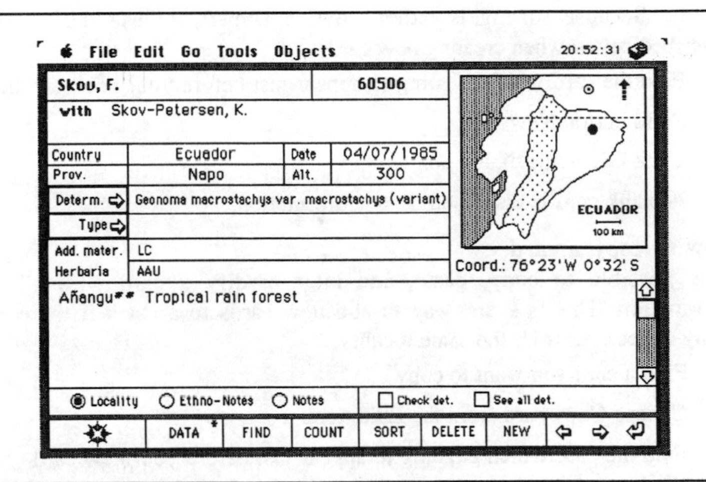

Figure 21. A card in the collections stacks.

9. A cross reference if the collection is type for a taxon
10. Other material to the collection (liquid, photos, slides *etc.*)
11. List of Herbaria where the collection is kept
12. Longitude (degrees)
13. Longitude (minutes)
14. Latitude (degrees)
15. Latitude (minutes)
16. Hemisphere (north or south)
17. Locality
18. Ethno-botanical notes
19. Other notes
20. List of previous determinations.

Adding and deleting cards

How to add a new card

New cards can be created anywhere in the stack. A new card is inserted after the currently displayed card. The stack can be sorted using the Sort button. Because sorting is rather slow in HyperCard use the method described below when creating new cards.

- Find the card which is going to appear just before your new record.
- Press button New
- Click OK in the dialog box
- A blank card is displayed. Enter information.

How to copy a card

It is possible to copy, paste, and later modify a card with all its information. This is a fast way to add new cards to a stack if there are many collections from the same locality:

- Find a card you want to copy
- Choose Copy card from the Edit menu
- Find the card which is going to appear just before your new card
- Choose Paste card from the Edit menu

- Make the necessary corrections to the new card.

How to delete a card

- Find the card you want to delete
- Press the Delete button
- Click OK in the dialog box.

Entering information

Main collector

The collections stack is sorted by this field and the number field. Use this field for the main collector only! If there are more than one collector, use the "Collected with" field to record their names.

> IMPORTANT: Type the last name first, initials afterwards, if you want HyperTaxonomy to sort your stack properly.

Example:
Skov, F.
Holm-Nielsen, L. B.

Collections Number

Use this field for the number of the collection. Mixed numbers like "345a " or "67876/3" are allowed. If a collection is without number, type *s.n.* in the field.

Other collectors

This field is for all other collectors mentioned on the label in addition to the main collector. If there are more than one they will be cited as *et al.* If there is only one co-collector he/she will be cited by name with the main collector. An example: If field Collector contains "Skov, F." and "Skov-Petersen, K." is in field "Collected with", the collection will be cited as Skov & Skov-Petersen.

Date

The correct format is day/month/year.
Examples:
Jan 6 1987 = 06/01/1987
Jan 1987 = 00/01/1987
1987 = 0/0/1987

If a label states that a collection is collected between Apr 10 and Apr 23, 1987 use 00/04/1987 to indicate the date as a month without further precision.

Country

The Neotropica version of HyperTaxonomy contains a list of countries from the region. Click Country to display the list, choose one and press Select to enter it to the field. Names of countries can also be typed in manually.

Province

Use this field for province or similar subdivision of the country. The Ecuador version of HyperTaxonomy has a list of all provinces in the country. Click Province to see the list (Fig. 22).

Altitude

Enter one value for altitude only. HyperTaxonomy does not work with ranges. Convert all values to meters before entry. Use numbers only (no text!) in this field.

Determination

The determination field constitutes a link to the taxonomy stack. This link is used to create lists of collections in the taxonomy stack. It is important that the spelling of a name is absolutely correct! HyperTaxonomy checks

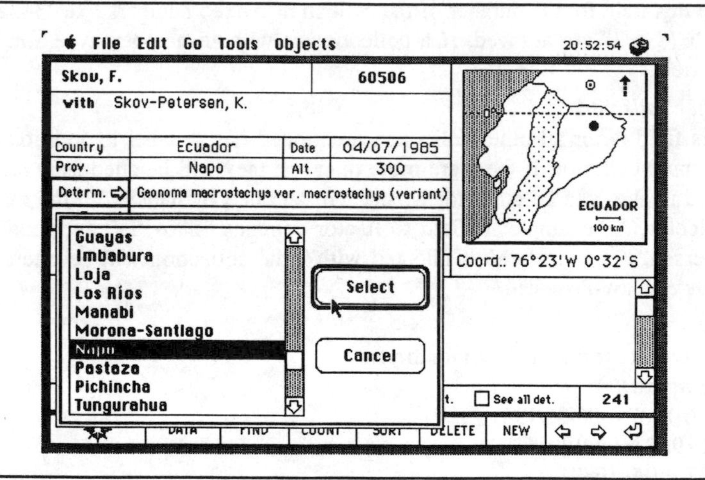

Figure 22. List of provinces.

the spelling this way:

- Click button Check Det
- HyperTaxonomy checks each determination against the taxonomy stack and informs the user if a suggested name is wrong or just missing. A "revert to old record" option is offered.

IMPORTANT: Remember to enter the rank of the taxon in brackets. The determination field can be used to determine collections to genus or subgenus if a species determination is uncertain.

Examples of allowed determinations:
Geonoma (genus)
Geonoma subgen. Geonoma (subgenus)
Geonoma maxima (species)
Geonoma pycnostachys var. stricta (variant).

HyperTaxonomy keeps a list of all determinations applied to a collection. Check button Old det. to see the list. This field can be edited and you can enter your own comments.

Cross reference if the collection is the type of a taxon
HyperTaxonomy supports four different types (and corresponding isotypes):

- Holotype
- Neotype
- Paratype
- Syntype.

The default setting is Holotype. If a specific collection is the holotype for a taxon, the Type field should contain the name of the taxon and its rank (in brackets). Is the specimen a neo-, para- or syntype, click left of the word Type, and select the desired type via dialog boxes. The new type will be displayed.

Other material belonging to the collection
This field is for a list of additional material belonging to a collection. This might be: liquid collections, photos, drawings, bulky material, pollen samples, *etc*. Abbreviations for most types of additional collections are convenient: (liquid collection = lc, photos = ph, *etc*.).

List of Herbaria from where the collections has been seen
Use herbarium acronyms according to Index Herbariorum and separate them by a comma.

Coordinates (fields 12–16)
The distribution map is restricted to a specific area depending on the version. The coordinates must be in the range of this area, *i.e.* the Neotropics: 35°–105°W, 23°N–23°S. There are five fields, two for latitude (degrees and minutes), two for longitude (degrees and minutes) and one for northern/ southern hemisphere. Type N or S in the latter, user numbers only for the four other fields.

In order to save screen space, only one of the following three fields is displayed. Field display is controlled with the radio buttons below.

Locality
This field is for locality data, including information about vegetation and habitat (Fig. 23).

In the final report, you may not want to cite all collections and all locality information about them. HyperTaxonomy makes it possible to choose exactly which collections to include and how much information to add to each:

Type the text to be included in the report in the beginning of the field. Put "##" after the last character (it is not necessary to include a final period). The string "##" tells HyperTaxonomy to include this card in the list of cited collections and to use the text before it as the locality. In the example shown in Fig. 23 only the word "Añangu" will appear in the report.

EthnoNotes
This field is for notes about ethno-botany like vernacular names and uses

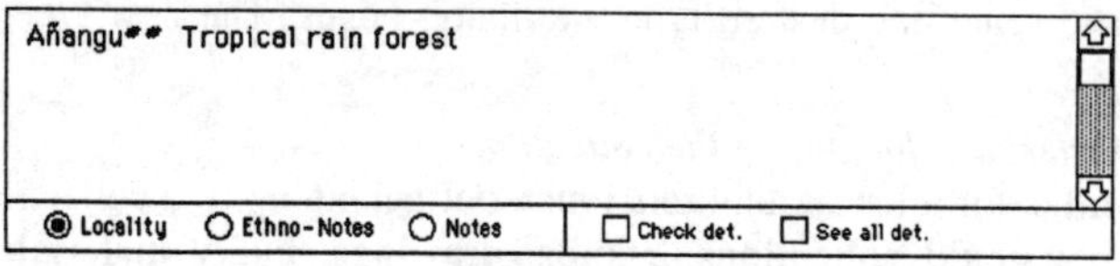

Añangu## Tropical rain forest

Locality Ethno-Notes Notes Check det. See all det.

Figure 23. Field for entering locality data. Note that only text before the string "##" will appear in the report.

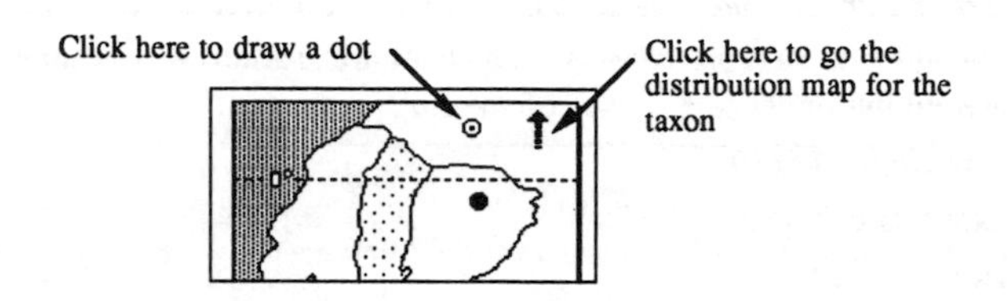

Figure 24. Buttons for drawing a distribution dot and for going to the maps stack.

of plants. No specific format or order of items is required.

Other notes
Use this field to enter all types of notes about a specimen. It is for small descriptions of species as sometimes found on labels or personal observations. The field can also be used for notes about the state of specimen, or for reminders like "this specimen has well preserved staminate flowers and could be used for pollen analysis".

Actions

How to create a dot map

The dot map in the upper right corner is redrawn by HyperTaxonomy when the button shown on the left is pressed (see Fig. 24). The location of the dot is determined by the geographic coordinates typed in fields 12–16.

How to add a new data sheet

If you want to add data to the data- or spreadsheet:

- Press the DATA button
- If a data sheet already exists for this collection it will be displayed
- Click OK if you want to create a new data sheet (see information about the measurements stack in chapter 8)
- A new card is created and automatically linked to the collection card.

> NOTE: If a determination of a collection is changed, the name on the data sheet will be changed as well.

Specimen cards with an existing data sheet are marked with a small asterisk in the upper left corner of the DATA button.

How to sort the stack

- Click button Sort.

How to find information

- Click button Find.

See chapter 2 for further details.

Count how many times a specific string occurs in the stack

- Press button COUNT
- Type a text string and HyperTaxonomy returns the number of cards on which the string occur.

Navigation

The figure below shows which stacks are accessible from the collections stack.

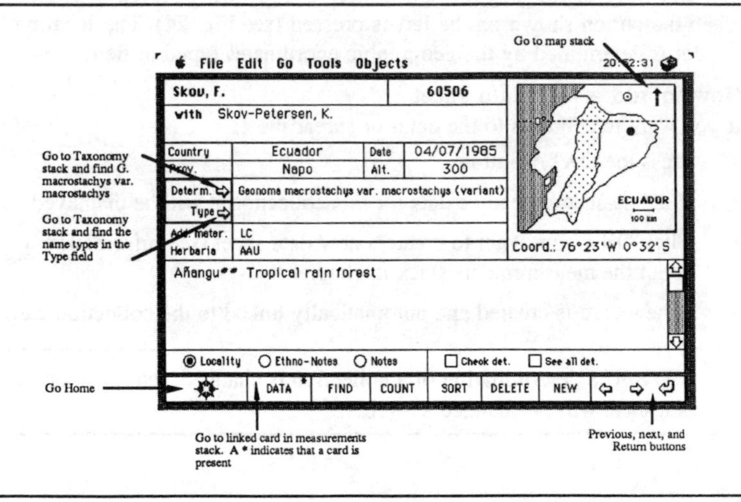

Figure 25. Collections stack - where to go from here.

6. LITERATURE STACK

This stack contains information about all literature involved in a taxonomic study. Each card holds information about one reference, *e.g.*, a book, a chapter in a book, or an article.

Card design

A card from the literature stack is shown below. It contains the following fields:

1. Author
2. Year of publication
3. Title of article or chapter in book
4. Periodical or book title
5. Volume number
6. Reference starting on page
7. Reference ending on page
8. Editors

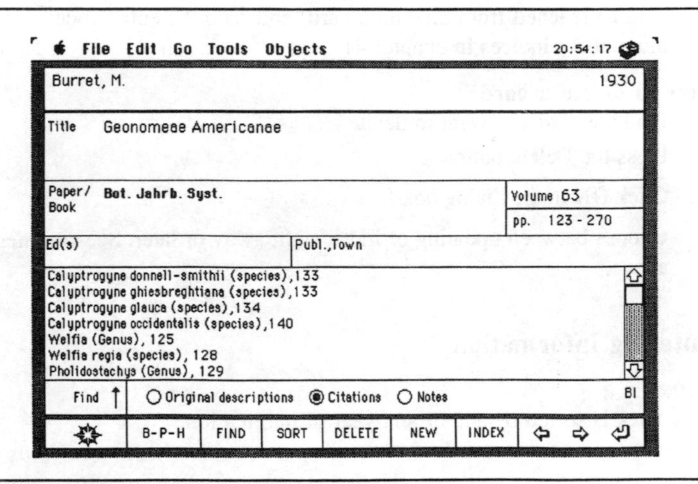

Figure 26. Card from the literature stack.

9. Publisher and place of publication
10. List of original descriptions of new taxa occurring in the reference
11. List of other taxa mentioned in the reference
12. Comments and notes about a reference
13. Bibliography.

Adding and deleting cards

How to add a new card

You can create a new card anywhere and sort the stack afterwards. It is, however, faster to insert the card in its correct place in the stack from the beginning.

- Find the card which is going to appear just before your new record
- Press button New
- Click OK in the dialog box
- A dialog box prompts for name of author and year of publication
- Another dialog box gives two options: immediate updating of the index or no update. Updating an index takes about 5 seconds. If you don't update the index right away, remember to go to card "updating" (can be reached from the home card) and redo the entire index (see more about indices in chapter 4).

How to delete a card

- Find the card you want to delete
- Press the Delete button
- Click OK in the dialog box
- Choose between updating of index right away or later. See comments above.

Entering information

Author
This stack is sorted by author and year of publication.

> IMPORTANT: Type the last name first, initials afterwards. Separate co-authors by a comma or "and" Do not use "&" which will cause trouble when the index is updated.

Examples
Burret, F.
Balslev, H. and Barfod, A.
Barfod, A., Henderson, A. and Balslev, H.

Year of publication
Always use a number in the beginning of this field, even if you don't know the actual year of publication. Entries like "?" or "xxx" are not allowed, but it is permitted to type "1897a" or "196?"

Title of article or chapter in book
Use this field for the title of the article or the title of a chapter in a book. If your reference is a book, use the next field for the book title. HyperTaxonomy must know whether your reference is a book or an article, and it recognizes articles or chapters by having both a title and a book/periodical title.

Name of book or name of periodical
Type the name of the book or the abbreviation of the periodical in this field. B-P-H abbreviations can be stored and searched for in the B-P-H stack (see chapter 10).

Volume number
This field is for volume number. You can type text or numbers.
Examples: 5, V, 2(3).

On pages (fields 6 and 7)
These two fields are for the first and last page number of the reference. If there is only one page, enter the number in the last field.

Editors
Use this field if the reference is a chapter in a book. Use the same rules for entering editor name(s) as for author.

Publisher and place of publication
This field is used for books only. Type the name of the publisher and the town in which the book is printed.
Examples: "Academic Press, New York" or "Collins, London & Glasgow"

To save screen space, only one of the three fields mentioned below is displayed. Three radio buttons control which field is shown.

> NOTE: The lists for original descriptions and citations are also used as indices to the taxonomy stack. Set the button Find to Write if you want to enter information.

List of original descriptions
Enter a list of all original descriptions (= descriptions of new taxa or transferring of taxa) found in the reference into this field. The field constitutes a link to the taxonomy stack and is where HyperCard searches for information about publications and publication date (see chapter 4). Use a line for each name. Separate name and information about the page on which it is found by a comma. Do not use space between items!

> IMPORTANT: Names of taxa must be followed by the rank in brackets after the name.

Example:
Geonoma (Genus),45
Geonoma acaulis (Species),45
Geonoma maxima (Species),46, pl. 34, Fig. 3

List of other cited taxa
This field is used for tracking citations of taxa which are not original descriptions. It is an easy way of remembering where a name has been mentioned. See how to enter data above.

Notes
This field is for additional notes of any kind. It can be used for key words, short abstracts, or any comments useful for later information retrieval.

Bibliographical notes
Short note about where to find a reference. Like BI = Botanical Institute library.

Actions

How to sort the stack

- Click button Sort.

How to find information

- Click button Find.

See chapter 2 for further details.

Navigation

The figure below shows which stacks are accessible from the literature stack. Note that fields Original Descriptions and Citations also function as indices to the taxonomy stack. Point at and click on the desired entry to see it work.

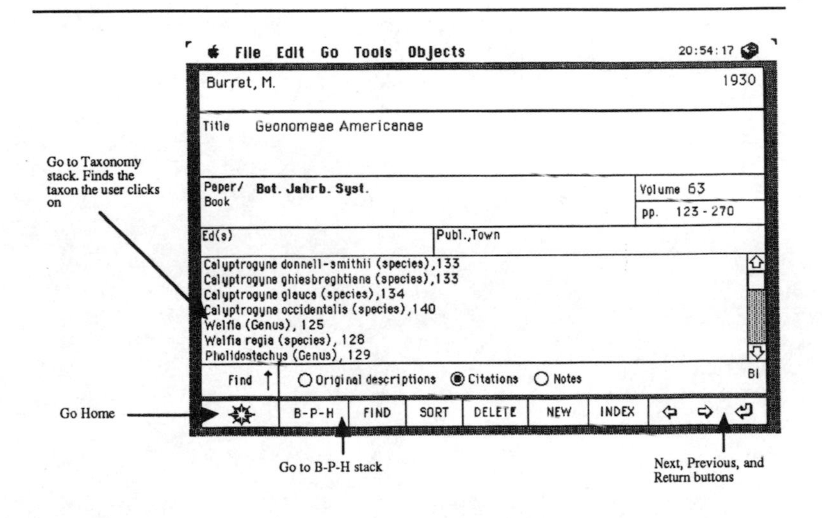

Figure 27. Literature stack - where to go from here.

7. TAXONOMY STACK

This stacks is, next to the home stack, the most important in HyperTaxonomy. It records every taxon relevant to a study, from variety to family. Most of the information is retrieved from the literature stack, the measurements stack, and the collections stack. HyperTaxonomy uses the taxon name and its rank to search for information in these stacks.

> IMPORTANT: The taxonomy stack has a hidden field where the name of the taxon and its rank in brackets is stored. The name and the rank is separated by one space only. This string must be used when a reference to a specific taxon is needed, *i.e.*, in the collections stack (determination and types), in the literature stack (original citations and citations), in the measurements stack (taxon names), and in the taxonomy stack itself (synonyms and taxonomic hierarchy). See the relevant stacks for further information.

Information retrieved from other stacks

Data from literature stack

- Full reference of original publication
- Year of first publication
- A list of literature references which mention the taxon.

Data from collections stack

- Data about type specimen(s)
- A list of all collections determined to a given taxon
- A list of notes, altitudinal range, and additional collections for a given taxon
- A list of geographical coordinates used for map drawing.

Data from measurements stack

- Raw descriptive morphological data used for species descriptions.

Data from B-P-H stack

- Author abbreviations.

Card design

A card from the taxonomy stack is shown in Fig. 28. It contains the following fields. Fields which are not filled in by the user are shown in italics:

1. Taxon name
2. Rank of taxon
3. Author abbreviation
4. Taxonomic status ("Category")
5. Year published
6. Correct name in case of a synonym
7. The next higher taxonomic level ("Part of")
8. *Full reference of the original description or where the taxon was transferred*
9. *Data about original collection (type species, or type genera)*
10. Data about neotype collection (if any)

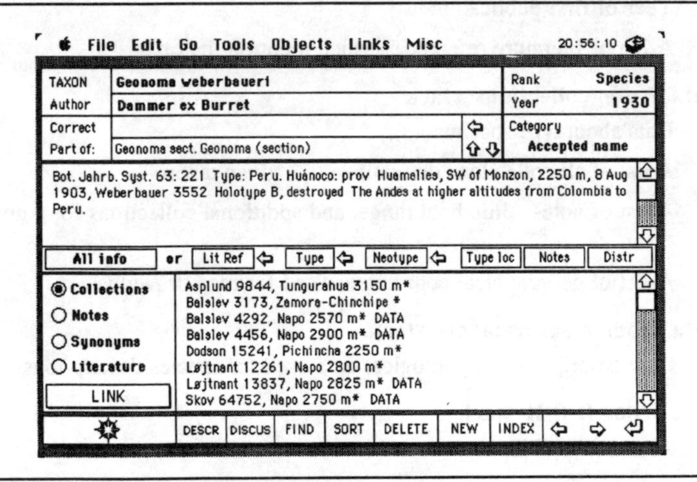

Figure 28. A card from the taxonomy stack.

11. Whereabouts of the types (for species and lower ranks only)
12. Nomenclatural notes
13. Distribution of the taxon.

> All information from fields 8—13 can be seen, but not edited, in field All Info.

14. *A list of collections of the taxon*
15. *Notes about a taxon*
16. *A list of synonyms of a given taxon*
17. *List of all literature references about a taxon*
18. Description of the taxon (press button Descr to see the field)
19. Discussion (press button Discus to see the field).

Adding and deleting cards

How to add a new card

New cards can be created anywhere in the stack. A new card is inserted after the currently displayed card. The stack can be sorted using the Sort button. Because sorting is rather slow in HyperCard, use this method for adding new cards:

- Find the card which is going to appear just before your new record
- Press button New and choose OK in the dialog box
- A list of available taxonomic ranks will be shown - select one.
- A dialog box prompts for the name of the new taxon
- If you want the index of plant names to be updated choose OK. If many new names are added it is faster to choose No, and update the index afterwards (see about updating of indices in chapter 4).

How to delete a card

- Find the card you want to delete
- Press the Delete button and click OK in the dialog box
- If you want the index to be updated choose OK in the dialog box.

Entering information (by the user)

Name of taxon

This field is filled in when a new card is created (see previous page). A name is the key to information about a taxon, be sure that the name and the rank of the taxon together form a unique entity! Do not use author abbreviations after the name.

Examples of valid names:
When the taxon is a genus: *Geonoma*
When the taxon is a subgenus: *Geonoma subgen. Geonoma*
When the taxon is a species: *Geonoma acaulis*
When the taxon is a variety: *Geonoma acaulis var. tapajotensis.*

Rank

Taxon name and rank are obligatory entries for each card. The rank of a taxon can be changed if the user points and clicks on the rank field. This action displays a box containing a list of available ranks. See the Fig. 29, right.

Taxonomic status ("Category")

This field is for the taxonomic status of a name. Click on the field and select one of the possibilities from the list (see Fig. 29, left).

> NOTE: If a taxon is both a heterotypical synonym for an accepted name and a basionym for other taxa, choose "Basionym" as the taxonomic status. HyperTaxonomy "knows" the name is a synonym if the field "correct name" is filled in.

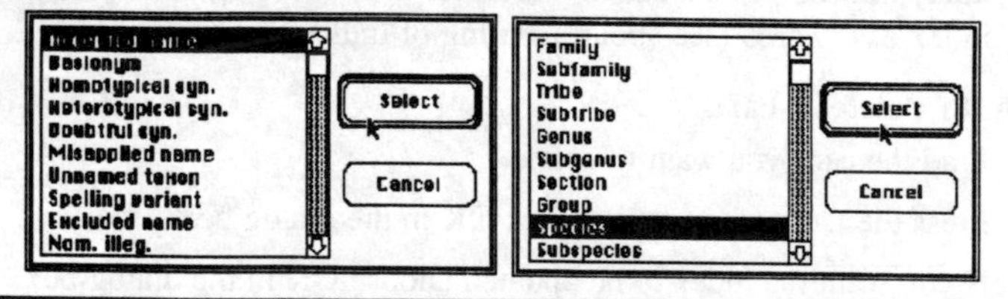

Figure 29. Lists of status and ranks available in HyperTaxonomy.

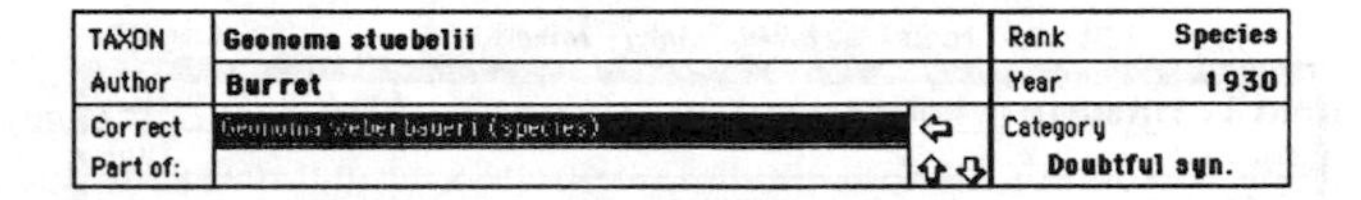

Figure 30. How to enter the correct name for a synonym.

Author (abbreviation)
Type the name of the author (or authors) in the accepted abbreviated form in this field. Correct abbreviations are found in the B-P-H stack. Choose Author abbr. from Misc menu (Chapter 10).

Year (of publication)
You can fill in this field or let HyperTaxonomy find the correct date in the literature stack (see p. 48).

Correct name (in case the name is a synonym)
Use this field for the correct name in case the taxon is a synonym. Type name of the accepted taxon and its rank in brackets (see Fig. 30)

> NOTE: All synonyms will be listed chronologically after the accepted name in the final report.

Hierarchy ("part of")
Use this field to indicate which higher taxon the current taxon belongs to. This feature is useful for linking related taxa. Type name and rank of the next higher taxon (see Fig. 31). Use the Up arrow right of the field to go up the hierarchy. In the example shown below, a card with information about *Geonoma sect. Geonoma* would be shown.

To go down the hierarchy use the down arrow. Clicking this button shows a field with all available subdivisions of a given taxon (Fig. 32).

TAXON	Geonoma weberbaueri	Rank	Species
Author	Dammer ex Burret	Year	1930
Correct		Category	
Part of:	Geonoma sect. Geonoma (section)		Accepted name

Figure 31. How to enter the taxon which the current taxon belongs to.

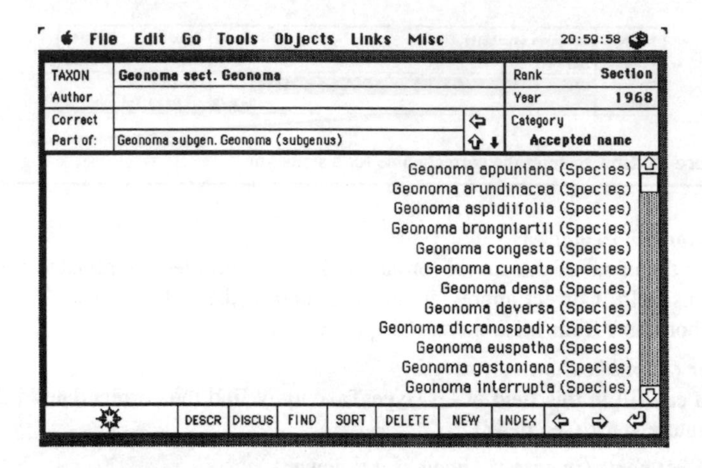

Figure 32. How to move down the hierarchy.

NOTE: To recalculate the list of subdivisions, press the command-key while clicking the down arrow.

Type Loc (whereabouts of the types)
This field is for information about where the types of a species or lower ranks are kept. Use the format recommended for the given project.
Example: Holotype B, destroyed; neotype G, n.v.; isoneotypes AAU, MO.

Notes (nomenclatural notes)
This field is for additional notes about the nomenclature of a taxon.

Dist (distribution of the taxon)
This field is useful when doing local floras for storing information about the distribution of a species outside the study area.

Actions

How to search for and retrieve information from linked stacks
HyperTaxonomy shares some features with relational databases. Retrieval of information from linked files (= stacks in HyperCard) is one of them.

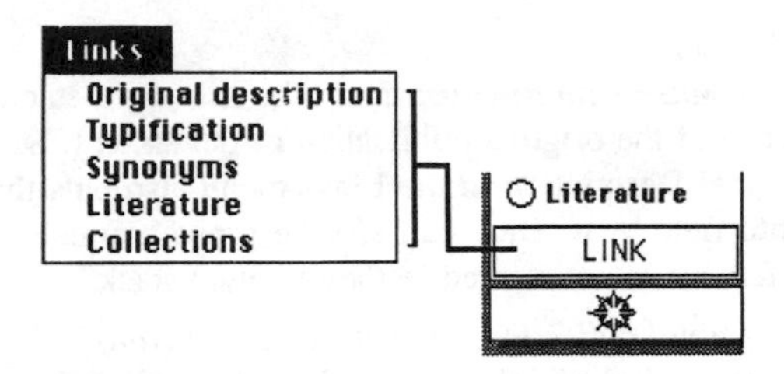

Figure 33. Links menu and the link button.

HyperTaxonomy, however, does not automatically retrieve the linked information. Updating of the stack is controlled from the home- or the taxonomy stack. In the taxonomy stack all retrieval of linked information is controlled by the Links menu and the LINK button found on the lower left (Fig. 33).

Select all the links you want to check, and press button LINK to see it work.

> NOTE: The LINK button works only if the correct information has been added to the appropriate stacks. (See literature-, collections-, and measurements stacks).

Field displaying all information

Information about a taxon gathered from different stacks are kept in separate fields: one for place of original description, one with information about a type, one containing nomenclatural notes, *etc*. It is often convenient to see all this information at the same time. Press button All Info to display a field containing such a summary (Fig. 28). This field can not be edited. To correct data, select the proper field, by choosing on of the buttons right

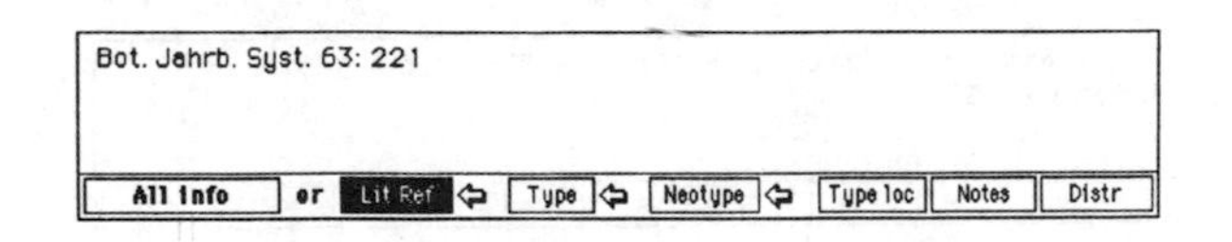

Figure 34. Field for literature reference.

of button All Info.

Literature reference
This field is for linked information from the literature stack. It contains the full reference of the original publication of the taxon (Fig. 34). Checking Original Description in the Links menu also puts the year of the description into field year. Data can also be typed manually if, for some reason, the reference is not wanted in the literature stack.

Typification (original collection, neotype, or type taxon)
These fields are for linked information from the collections stack or for information entered manually. They contain full citation of the original collection or a neotype collection if the original collection is destroyed. To make it work, add all data about the collection to the collections stack and apply the name of the taxon to the Type- or Neotype field. If you do not want to add the specimen to you collections stack, simply type the information to the field manually (Fig. 35).

If the taxon is above species rank and is typified with a species or another taxon, use this example as a model: If *Geonoma simplicifrons Willd.* is type species for the genus *Geonoma*. Type "Lectotype: *G. simplicifrons Willd.*" in the original collection field.

> NOTE: If a taxon has a basionym the type collection must be linked to that name. To indicate where to find the type collection, type an asterisk followed by the name of the basionym and its rank in brackets in the Type field.

Examples
Geonoma maxima (Poit.) Kunth is based on *Gynestum maximum Poit.* Type: "**Gynestum maximum* (species)" into the type field. This tells HyperTaxonomy that *Geonoma maxima* is based on *Gynestum maximum* and therefore on the type collection of this taxon.

Figure 35. Field for type reference.

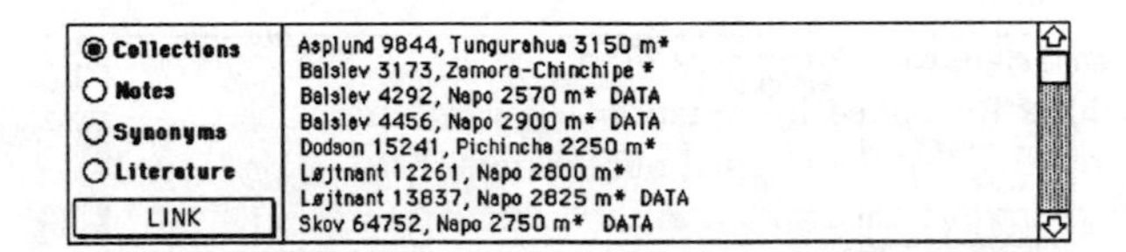

Figure 36. A list of collections determined to a given taxon.

A list of collections of a species

This field is for linked information from the collections stack. It contains a complete list of all the collections determined to the given taxon (Fig. 36). The list is also an index to the collections stack. Point and click on a collection to go to that card in the collections stack. Province and altitude are included to give an impression of the distribution. If collections are marked with an asterisk, it indicates that they are destined for the final report and have been marked with "##" in the locality field (see chapter 5). "DATA" means that a collections has a card in the measurements stack.

A list of notes from collections card

This field is also for linked information from the collections stack. It contains various types of information:

- Number of collections of a taxon
- Its altitudinal limits
- A list of all additional collections of a taxon
- All notes from field notes on the collections cards
- All ethno-botanical notes.

This field is especially useful for writing the final description and discussion (see pp. 52-53).

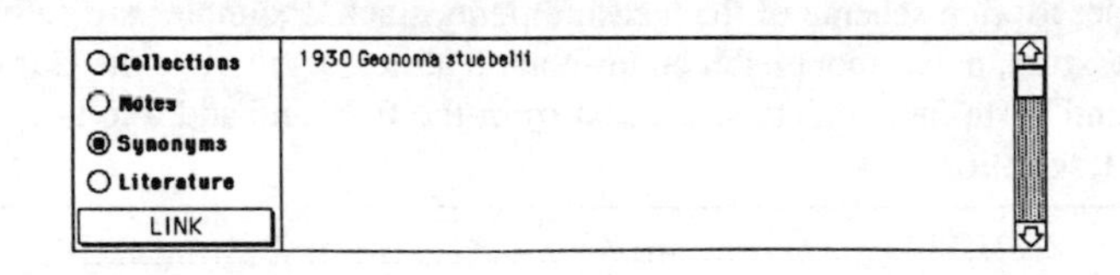

Figure 37. List of synonyms to a given taxon.

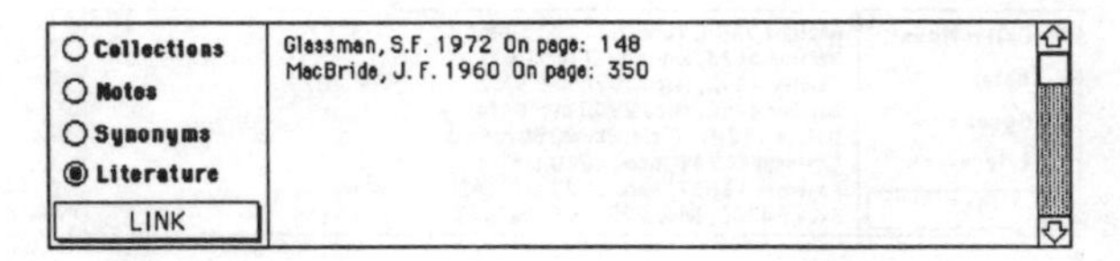

Figure 38. A list of literature references, where the taxon has been cited.

List of synonyms

This field is for linked information from the taxonomy stack. It contains all synonyms of a taxon. The field works as an index to synonyms.

Point and click at a desired name, and the card in the taxonomy stack containing information about it will be displayed (Fig. 37).

List of literature references

This field is for linked information from the literature stack. It contains a list of all references which mention the taxon. It also works as an index to the literature stack. Point and click at a desired name, and the corresponding card in the literature stack will be shown (Fig. 38).

Working with information (refined description and discussion)

Description

Descriptions made by HyperTaxonomy are raw sketches only, but can be used as backbones for more refined descriptions. A special field for such an expanded description is available by pressing the button Descr.

The first step is to copy the description from the measurements stack. Click button Descr. while pressing the command-Key or go to the measurements stack, select and copy the information, go back and paste into the Descr. field. Comments and corrections can now be made (Fig. 39).

> NOTE: The note field is very useful during descriptional work. This field contains a lot of information which does not fit the more rigid description scheme of the measurements stack. Examples are colours, texture, notes about habit ethno-botanical notes, *etc.* Use the Copy and Paste functions to select text from the field and add it to the description.

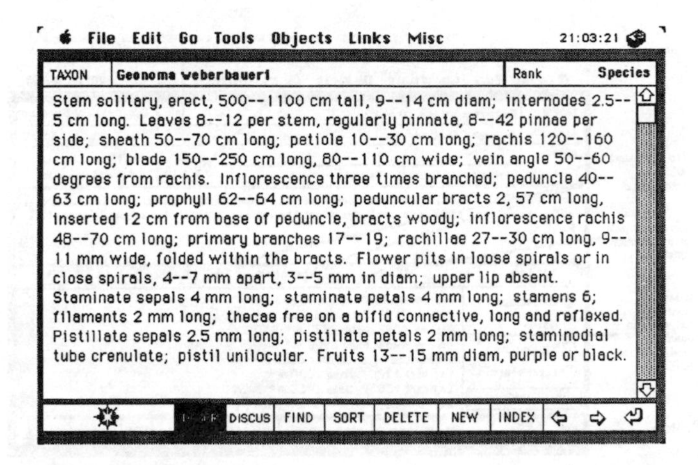

Figure 39. Description field.

Discussion
Discussions usually follow a description in a taxonomic treatment. Use this field to edit and enter the discussion.

Navigation

The figure below shows which stacks are accessible from the taxonomy stack. Note that some fields also function as indices to the collections- and the literature stacks. Point at and click on the desired entry with the mouse to see it work.

Navigating with the MISC menu

Use the MISC menu to go to:

- A distribution map (see chapter 9)
- The spreadsheet in the measurements stack with morphological data about a taxon (see chapter 8)
- The B-P-H stack with commonly used abbreviations of authors (see chapter 10).

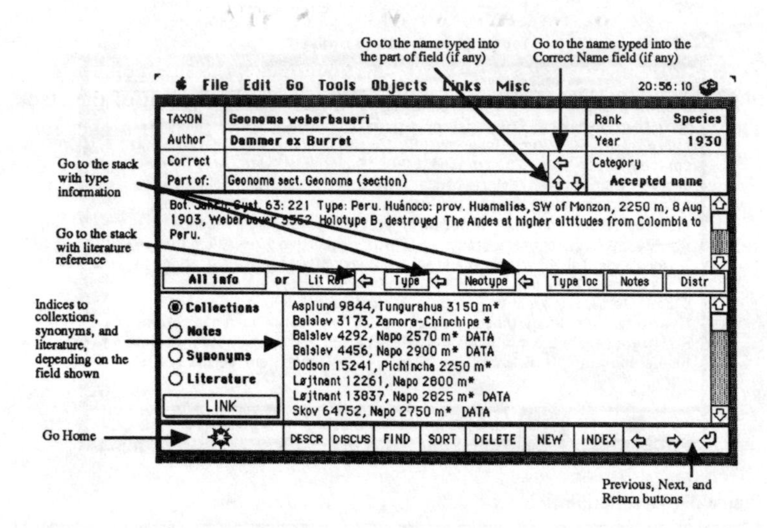

Figure 40. Taxonomy stack - where to go.

8. MEASUREMENTS STACK

The purpose of this stack is to store and manipulate measurements of morphological data. Descriptions of collections are the basic unit of the stack. They can be combined into descriptions of species or taxa of higher rank. The data format follows the DELTA format of Dallwitz (1980) and supports up to 45 characters. Characters can be quantitative or qualitative. Quantitative and ordered multi–state characters permit minimum and a maximum values to be entered. All character states applying to a taxon or a collection can be entered for unordered qualitative data.

Card design

The design of a card is shown in Fig. 41.

Note the small window in the lower right corner. Information about character "Habit" is displayed by clicking button Habit in the upper left corner. All characters can be viewed and edited this way.

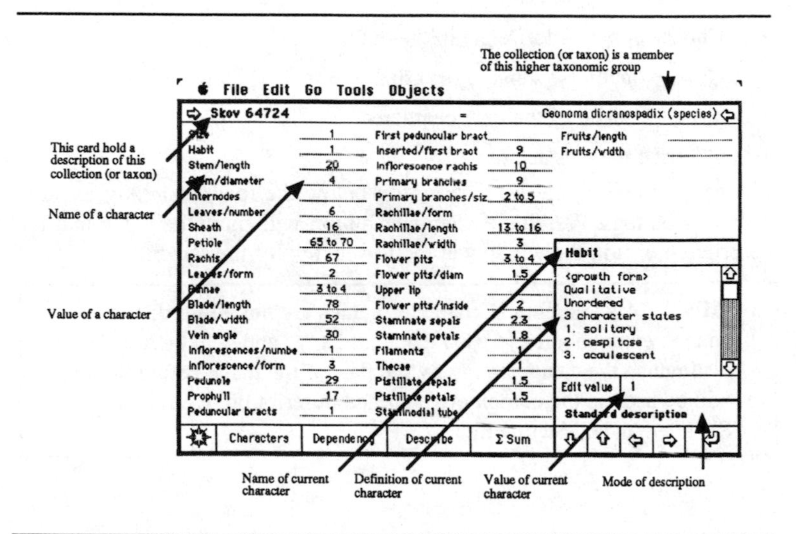

Figure 41. A card from the measurements stack.

Working with characters

How to set up data for taxon identification and description is explained in detail by Pankhurst (1986). Taxa are described by a set of characters and the states of these characters. Characters are either quantitative or qualitative.

Quantitative characters have an indefinite number of character states. A typical example could be stem length or number of branches of an inflorescence. Quantitative characters are either whole numbers (counts or integers) or fractional (real numbers). They must be assigned key states which divide the character into a definite number of states in order to be of any use in, *e.g.*, key construction.

Qualitative data have a definite number of character states. An example could be stem colour (brown, green, yellow, or reddish) or leaf form (entire, trijugate, or pinnate). Qualitative characters are ordered or unordered. Ordered characters can be described in ranges, *e.g.*, a leaf form can be entire to pinnate. Some character states rule out other characters. If a character such as inflorescence form has state "spicate", it is meaningless to describe the number and size of the branches.

Adding new characters

- Press button Characters
- Choose option Add/Delete in the dialog box
- Choose option New in the next dialog box
- Type the name of the new character
- Describe the character.

The new character appears as a button below the row of already existing character buttons. Pressing a character button will show the name and the description of it in the window at the lower right of the screen.

> IMPORTANT: Identical character names are not allowed. If you wish to enter an already used name add a "/" and a text string to distinguish the characters. Only the part of the name before the dash will be used in the computer generated description. Example: Stem/length and Stem/width.

IMPORTANT: Character names ares used in descriptions. Read this part of the manual carefully before constructing a morphological database to obtain the best possible descriptions.

Deleting characters

- Press button Characters
- Choose option Delete in the dialog box
- Choose the desired character in the shown list.

WARNING: Deleting a character cannot be undone.

Reorganizing characters on the card

The order of characters on the screen can be changed:

- Click button Characters
- Choose Reorganize
- Select the character which position you want to change
- Select the character which should appear just before it.

Changing the name of a character

All character settings can be changed

- Choose the character to be changed
- Click on the character name (above in the box on the right, see Fig. 42)
- Change the name and click OK.

Changing the description of a character

- Choose the character to be changed
- Click on the exact line to be changed in the character editing window (see Fig. 42)
- Change the line.

WARNING: When a character is changed some of the existing data could be out of range. Be careful to change all data if necessary.

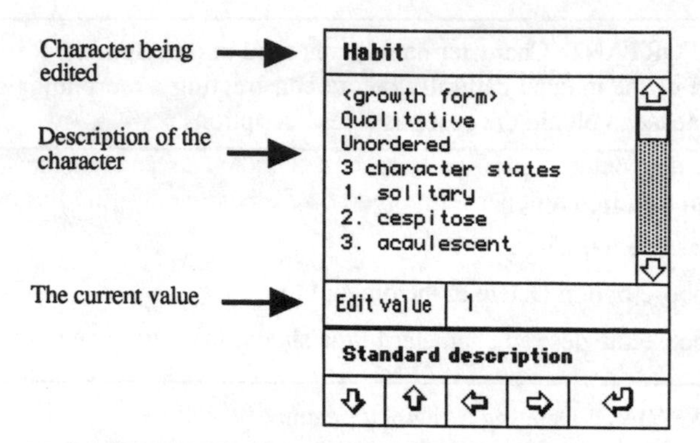

Figure 42. The character editor window.

Entering information

Click on the character for which you want to enter data. Look at the window at the lower right where information about the character is displayed. This window informs about type of character and acceptable states (Fig. 42).

The field name is shown above. Below the field name follows a description of the character. The current value of the character is shown. There are two ways to enter or edit values:

1. Press button Edit Value. HyperTaxonomy will ask you for maximum and minimum value or you will be prompted for each of the possible states (Fig. 43).
2. Enter the value directly in the field to the right of the character name.

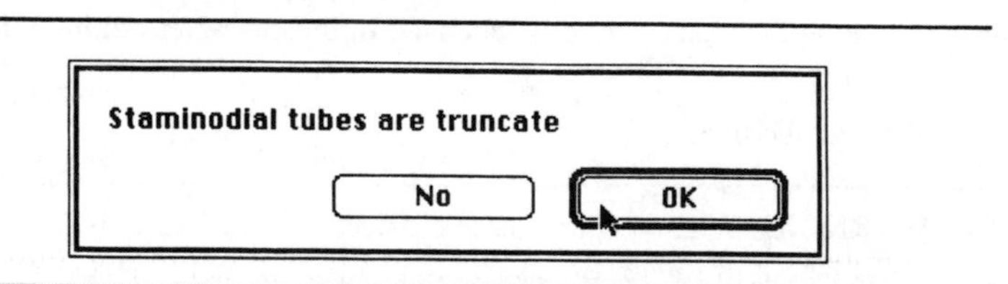

Figure 43. Dialog box prompting the user for a character state.

Working with morphological data

Accumulating data

The first data entered are usually specimen data. The name of the specimen will be placed in the upper left corner. There is an empty field in the upper right corner of the card for the next higher taxonomic level. Typically a specimen will be assigned to a species. When several sets of morphological data about specimens of one species have been entered, press the Up arrow to see the species card. All morphological data for this taxon can be accumulated by pressing button ΣSUM. This process can be continued to the highest taxonomic level. Species can be assigned to subgenera which in their turn are members of genera, *etc*.

> NOTE: Specimens are not the only source of morphological data. Descriptions found in books or articles can also be used. Instead of collector and number type the reference in the field on the top left.

Describing taxonomic entities

All taxa (and collections) can be described in a text format set by the user.

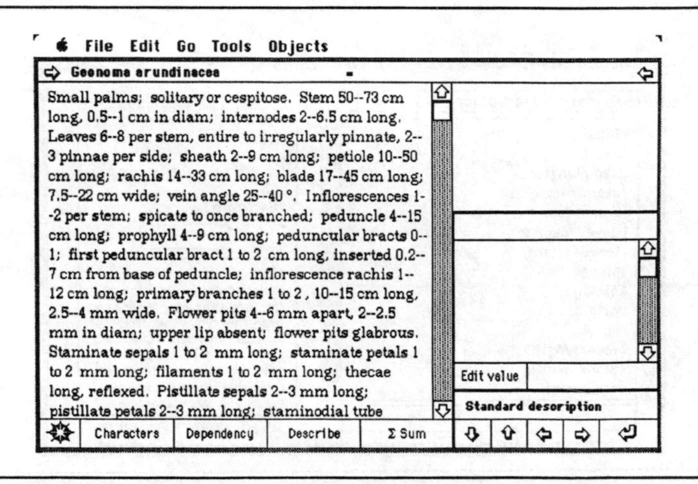

Figure 44. A card from the measurements stack with description field shown.

Each card has a field where the description can be seen. To see the description field press button Describe while pressing the command-Key at the same time. Use the same procedure for hiding the field. The field is shown in Fig. 44.

Note the field just above the navigation buttons on the figure. It says "Standard description". It can be changed and a user can make as many formats for description as necessary. To edit a description, click on "Standard description". This action brings you to the:

Description editor

The name of the card is shown in the field above. Below this field is a list of characters included in this description type. It is not necessary to include all available characters in a description.

The description editor gives the user several choices when creating a description:

- Include a character name in the description or not. A "<" before a character name tells HyperTaxonomy not to use the name in the description. (Note that character name is the part of a name before a "/").

It is not always desirable to use the character name in a description. In the example shown in Fig. 45, character name is used for stem/length but not

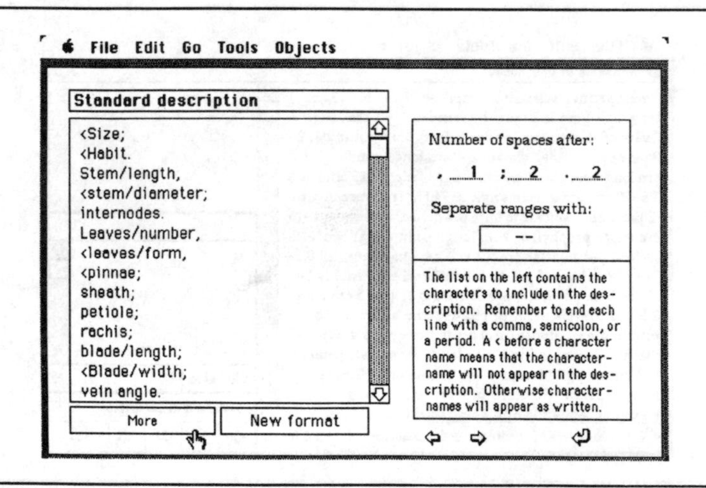

Figure 45. Card from the measurements stack showing the description editor.

for stem/diameter. The result will be a description reading: "Stem 50--100 cm long, 1--3 cm in diameter".

- Capitalize first character in the character name or not.

Names will appear as they are written in the list. Full stops are followed by capital letters.

- Which sign should be used to separate this character from the next.

Three types of separator signs are used in HyperTaxonomy:
comma, semi-colon, and period. Use the one you find appropriate. You can decide how many empty spaces to follow each of the different signs by changing the number typed in the three fields shown in Fig. 45.
Different ways of separating ranges can also be chosen here. Choose "4-5","4--5", or "4 to 5" depending on the type of manuscript you are preparing.

To add more characters to the list, press button More. It will display a list of available characters, and prompt you for the relevant questions.

A faster way of creating a new description format is to press button More while pressing the option-key. This action makes a complete list of all characters separated by periods. You can then manually edit this list to your purpose.

> WARNING: Never change the spelling of the characters or leave out a separator sign. HyperTaxonomy cannot construct a description without them.

> IMPORTANT: It is important to have the descriptions in mind when constructing character names and character states.

If a description should look like this: "stem 4--5 cm long", set the unit of the character to "cm long". If you want a description like this "Peduncular bracts 1 to 2, inserted 4 to 6 cm from the base", set the name of second character to "inserted/peduncular bract" and the unit to "cm from the base".

Press the return button on the lower right to go back.

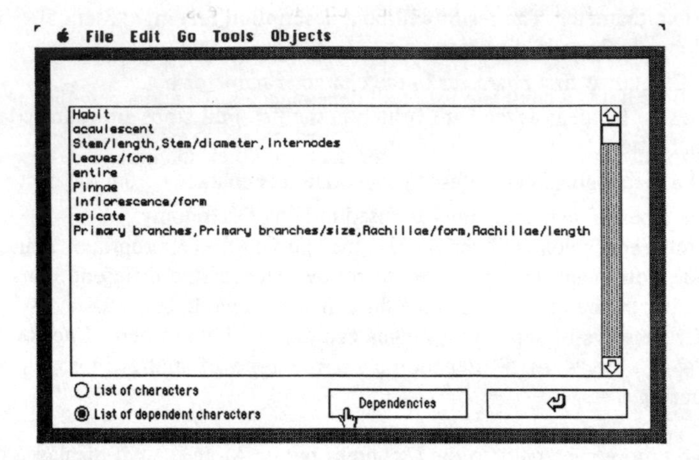

Figure 46. Card from the measurement stack with information about dependent characters.

Dependent characters

Some characters are dependent on others. If for instance a leaf blade is entire there is no sense in describing the number and length of the pinnae. HyperTaxonomy needs information about all dependent characters in order to create proper descriptions.

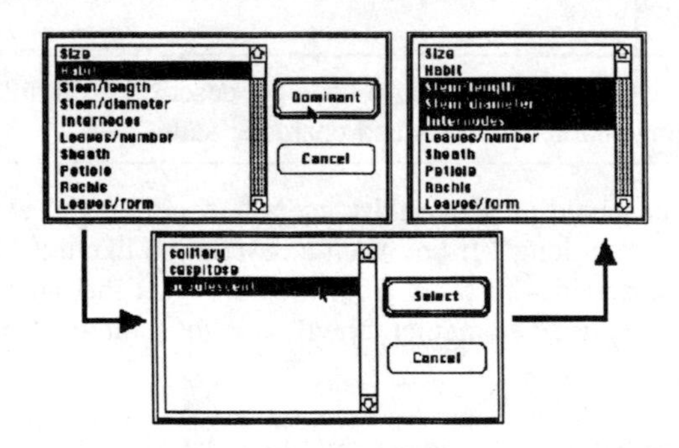

Figure 47. Sequence of dialog boxes, showing how to set dependencies.

To construct a list of dependent characters press button Dependency. The card shown in Fig. 46 will be displayed.

The field holds a complete list of all dependencies. To add a new one, press button Dependencies. A list of all characters will be displayed. Choose a dominant character. In this example <habit>. Click Dominant. A new list with all the states of character <habit> will be shown. Choose the dominant state "acaulescent". A third list containing all the characters will be shown. Select all characters which are impossible when character habit is acaulescent, *i.e.*, characters describing the stem. More than one character can be chosen by pressing the shift key while clicking on the desired character/s. Click OK when finished. The new instruction will be pasted in the field. Delete instructions by selecting them and use Cut from the Edit menu. The whole sequence is shown in Fig. 47.

Navigation

The figure below shows accessible stacks and cards.

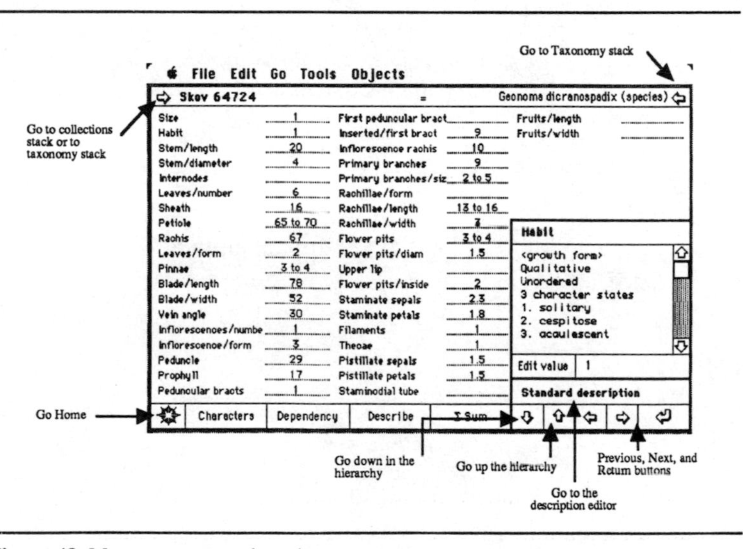

Figure 48. Measurements stack - where to go.

9. MAPS STACK

The map stack contains distribution maps for each taxon with a list of collections attached. The distribution maps can be printed or used as graphics in other Macintosh applications. Print quality is not sufficient for publishing, but these maps can act as templates for more elaborate distribution maps.

Card design

The design of a card is shown in Fig. 49.

Adding and deleting cards

New cards are created from the taxonomy stacks when "Distribution map" is chosen from the Misc menu. Cards can be deleted using the Delete card function in the Edit menu.

Actions

How to redraw a distribution map

If a distribution map has to be redrawn, there are two possibilities:

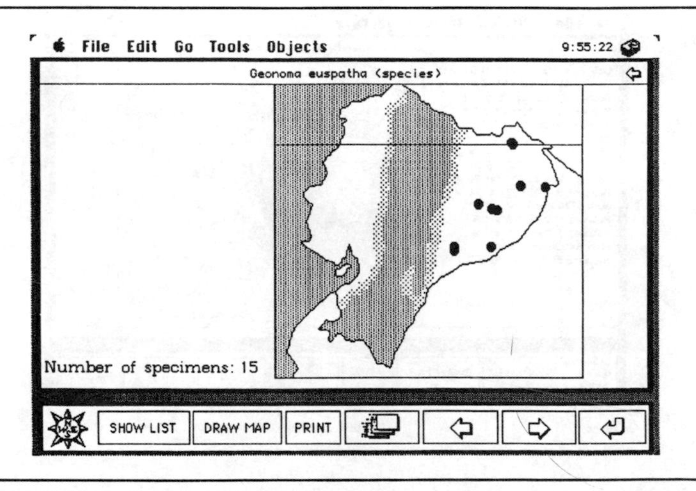

Figure 49. A distribution card from the maps stack.

- Click button Draw map
- Click No in the dialog box, if you don't want HyperTaxonomy to scan the collections stack for new coordinates

  or
- Click OK in the dialog box, if you want that search to be conducted.

Adding drawings and text to a distribution map

- Drag the toolbox to a corner of the screen and select the desired tool. Choose the Browse tool when you finish.

Print a map

Press button Print in order to hard copy the distribution map.

Navigation

The figure below shows which stacks are accessible from the maps stack. If you want to use the index shown in the figure, click button Show list.

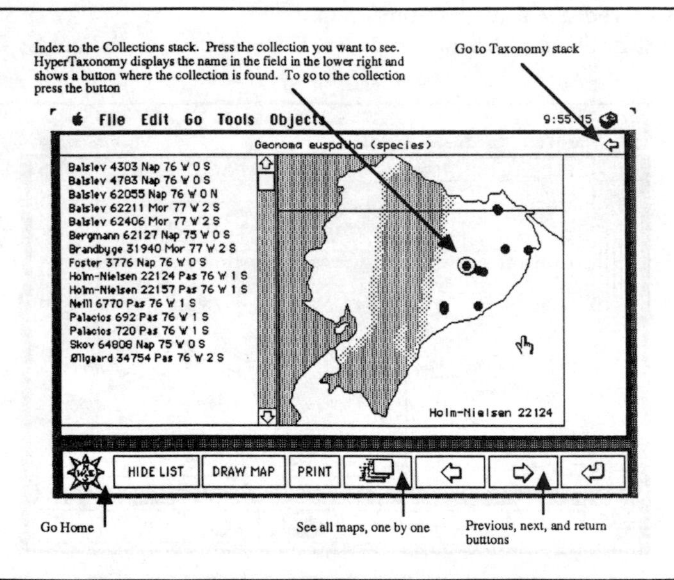

Figure 50. Maps stack - where to go.

10. B-P-H STACK

This stack provides a user of HyperTaxonomy with on-line help. Taxonomists use abbreviations of periodicals, author names, and books. Use this stack to store all abbreviations used in your work.

There are two backgrounds in this stack. One for Author abbreviations and one for B-P-H abbreviations. Fig. 51 shows a B-P-H card, Fig. 52 shows the background for the Author abbreviation index.

Card design

The design of a card is shown below.

Adding and deleting cards

How to add a new card

- Press button New.

How to delete a card

- Find the card to delete
- Choose Delete card from the Edit menu.

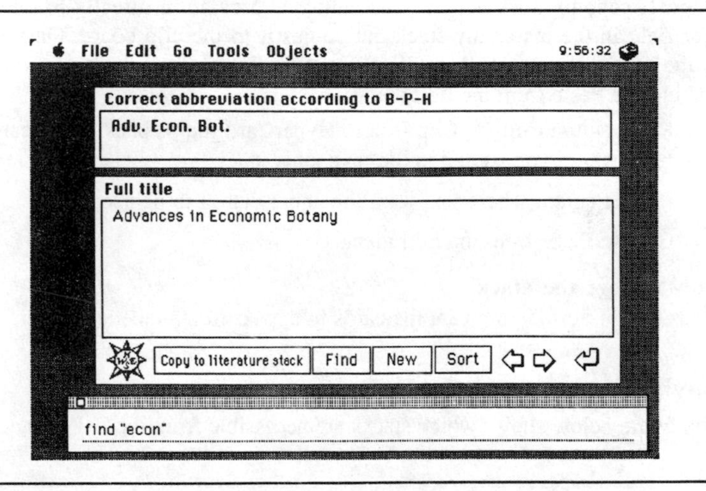

Figure 51. Card from the B-P-H stack with B-P-H abbreviations.

Entering information
Use only one abbreviation for each periodical. HyperTaxonomy uses the following works to standardize abbreviations: B-P-H, Kew Index and TL2.

Actions

How to find the desired information
When the stack is opened, the message box with the Find command is shown. Simply type part of the author name or the periodical you want to find and press return. If a wrong card is shown press return until you find the right one. If you cannot find a name in the stack, it must be entered.

How to transfer information
To the literature stack

- Press button Copy to Literature Stack. HyperTaxonomy will jump back to the card from where you started in the literature stack
- Press OK when the dialog box asks if you want to use the abbreviation
- The abbreviation is pasted into the proper field.

To the taxonomy stack
Author names are often composed of more than one name. For that reason, HyperTaxonomy does not copy an author abbreviation directly to the author field in the taxonomy stack but copies it to the clip board. On return to the taxonomy stack, place the cursor where the abbreviation is wanted and choose Paste from the Edit menu.

- Press button Copy to Clip Board. HyperCard jumps back to the card from where you started in the taxonomy stack
- Put the cursor where you want the abbreviation to be used
- Choose Paste from the Edit menu.

How to sort the stack
Press button Sort if you want the cards to appear in alphabetical order.

Navigation
The figure below shows which stacks are accessible from the B-P-H stack.

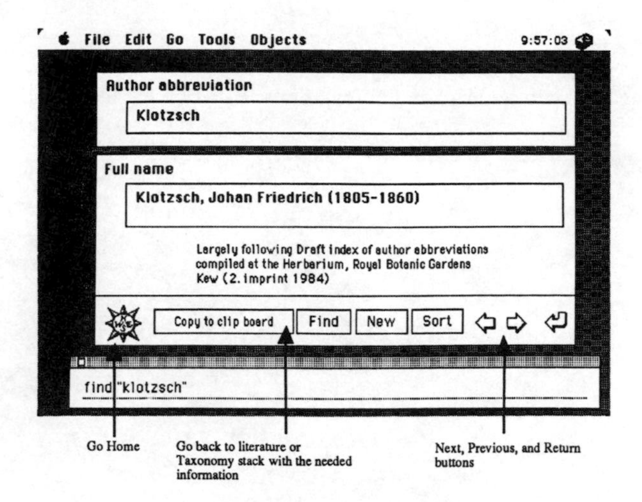

Figure 52. Card from the B-P-H stack with author abbreviations.

11. LITERATURE CITED

Allkin, R. 1989. Taxonomically intelligent database programs. — pp. 315–331. In: Hawksworth, D. L. (Ed.), Prospects in Systematics. Clarendon Press, Oxford.

Allkin, R. and Bisby, F. A. 1988. The structure of monographic databases. — Taxon 37: 756–763.

Beaman, J. H. and Regalado, J. C. 1989. Development and management of a microcomputer specimen-oriented database for the flora of mount Kinabalu. — Taxon 38: 27–42.

Dallwitz, M. J. 1980. A general system for coding taxonomic descriptions. — Taxon 29: 41–46.

Gómez-Pompa, A., Moreno, N. P., Sosa, V., Giddings, L. and Soto, M. 1985. Flora of Veracruz project: an update on database management of collections and related information. — Taxon 34: 645–653.

Goodman, D. 1987. The complete HyperCard handbook. — Bantam Books, Toronto.

Goodman, D. 1988. HyperCard developer's guide. — Bantam Books, Toronto.

Leenhouts, P. W. 1968. A guide to the practice of herbarium taxonomy. — Regnum Vegetabile 58: 1–60.

Pankhurst, R. 1986. A package of computer programs for handling taxonomic databases. — Computer Applications in the Bioscience 2: 33–39.

Pankhurst, R. 1988. Database design for monographs and floras. — Taxon 37: 733–746.

Shafer, D. 1988. HyperTalk™ Programming. — Hayden Books, Indianapolis.

Skov, F. 1989. HyperTaxonomy - a new computer tool for revisional work. — Taxon 38: 582–590.

11. LITERATURE CITED